A

*Poe's Heart and the Mountain Climber*

Also by Richard Restak, M.D.

*Mysteries of the Mind*

*Mozart's Brain and the Fighter Pilot*

*The New Brain*

# Poe's Heart

and

## RICHARD RESTAK, M.D.

HARMONY BOOKS • NEW YORK

the

# Mountain
# Climber

## EXPLORING THE EFFECT
## OF ANXIETY ON OUR BRAINS
## AND OUR CULTURE

Grateful acknowledgment is made to Random House Reference and Information Publishing for permission to use "Ninety-Nine Phobias" from *Grow Your Vocabulary by Learning the Roots of English Words* by Robert Schleifer. Copyright © 1995 by Robert Schleifer. Reprinted by permission of Random House Reference and Information Publishing, a division of Random House Inc.

Published by Harmony Books, New York, New York.
Member of the Crown Publishing Group,
a division of Random House, Inc.
www.crownpublishing.com

HARMONY BOOKS is a registered trademark and the
Harmony Books colophon is a trademark of Random House, Inc.

Printed in the United States of America

Design by Leonard W. Henderson

Library of Congress Cataloging-in-Publication Data

Restak, Richard M., 1942–
Poe's heart and the mountain climber : exploring the effect of anxiety on
our brains and our culture / Richard Restak.—1st ed.
1. Anxiety.    2. Anxiety disorders.    I. Title.
QP401.R47 2004
152.4'6—dc22            2004019309

ISBN 1-4000-4850-8

10 9 8 7 6 5 4 3 2 1

First Edition

*To Alison*

# Acknowledgments

GRATEFUL THANKS TO the following neuroscientists, psychiatrists, psychologists, philosophers, anthropologists, and historians who have helped with suggestions, discussions, interviews, useful papers, and criticism: Mohan Advani, David Amaral, Mark Barad, Michael Davis, Naomi Eisenberger, Frank Farley, Richard Greenberg, Edna H. Hong, Howard V. Hong, Kevin S. LaBar, Richard J. McNally, Kathleen Ries Merikangas, Jaak Panksepp, Elizabeth Phelps, Gregory Quirk, David V. Sheehan, Ben Shephard, Charles Spielberger, Dan Tranel, Lewis Winkler, John Wylie, Rachel Yehuda, and Allan Young.

*Learning to know anxiety is an adventure*
*which every man has to affront if he would not go to perdition*
*either by not having known anxiety or by sinking under it.*
*He therefore who has learned rightly to be in anxiety*
*has learned the most important thing.*
— SØREN KIERKEGAARD, 1844

# Contents

*Poe's Heart and the Mountain Climber*

# Introduction

AILY LIFE IN TWENTY-FIRST-CENTURY America provides a steady diet of anxiety-provoking events. On more days than we care to count, we awaken to television newscasts of riots, violent protests, and killings occurring at various locations throughout the world. Over breakfast, we read about infectious diseases that might evolve into a modern-day plague; additional power outages resulting from our outmoded energy delivery systems; computer worms and viruses with the potential to destroy our most secure databases; and escalating levels of drug use and violence both in large cities and small towns.

And it gets even more up close and personal. Upon our arrival at the office, we learn that the CEO has made yet another change in employee health insurance coverage, thus requiring us to choose yet another new doctor. While driving home from work, we flip from station to station on the car radio in search of information about potential terrorist threats, all the while reviewing in our mind various disaster scenarios that could conceivably befall us.

Anxiety has become such an integral part of our lives that Americans reported higher levels of anxiety in the 1990s than they did in the 1950s, the so-called Age of Anxiety. Among children, the situation is even worse. Starting in the 1980s, normal children experienced higher anxiety levels than adult psychiatric patients in the 1950s!

Media pundits and gurus have cited several possible reasons for this increase in anxiety. For one, near instant communication technology provides us with vivid video depictions of anxiety-provoking events occurring thousands of miles away, events that otherwise often bear little relevance to anything happening or likely to happen in our own lives. What's more, government officials, media marketers, and even scientists have learned an important principle: If they want to get our attention, they have to arouse our anxiety—if you doubt this, just watch the nightly newscasts or read your morning paper. Over the last decade or so, they've discovered (principally via the process of trial and error) that most of us pay more attention to those who speak to us of the terrible things that may happen than we do to people who assure us that everything is all right.

As a result of these converging influences, we're now exposed to information about innumerable nerve-racking calamities that might occur. To make matters worse, anxiety tends to be a cumulative emotion: If we become anxious about something today, then our anxiety will resurface whenever we encounter that same event or situation in the future. And since each day provides any number of anxiety-provoking events, the triggers for anxiety arousal increase over the years.

In response to escalating personal and communal anxiety, increasing numbers of us are falling prey to anxiety-associated illnesses. At the moment, more than 19 million Americans suffer from some form of anxiety dysfunction. Ask any primary care doctor and he or she will tell you that anxiety is the underlying cause for the majority of patient complaints. Because we are collectively feeling increasingly threatened, vulnerable, and helpless—that our lives are determined for us by forces outside of our control—our individual and community anxiety levels are on the increase. In response, we take various attitudes toward managing that anxiety.

For example, consider an article I encountered in the *New York Times* entitled "A Nice Place to Live if You Can Live with Terror." It described the attitudes of several wealthy Colombians living in Bogotá in February 2003, two days after terrorists detonated a 330-pound bomb in the parking garage of an exclusive sport and social club, killing 32 people and wounding at least 160. The statements of several of the interviewees are typical of different responses to anxiety, in this case the anxiety provoked by the uncertainty and risk associated with living under threats of terrorism.

"Six children died in there; how can that be?" asked the father of a thirteen-year-old who often played squash or miniature golf in the bombed structure. "It is absurd, so absurd."

"This march today is a march of fury for the loss of our countryman," shouted a famous Colombian actor during a demonstration protesting the bombing. "If they want to stop us they will have to kill all 40 million Colombians."

"Now all of us are worried because we could be in their sights," commented a retired surgeon while on a putting green at his Bogotá golf club.

"I will do the same things I did before but I will be more

careful," said a retired engineer. "I will not go to risky places, to certain restaurants, or take long car trips, or go to shopping malls on certain days."

"This is still a nice place to live," responded another retired engineer. "If you take precautions, you can live very, very well. You cannot just abandon that."

Notice the progression of attitudes expressed by these quotes (which I've presented in a different sequence than in the *Times* article): disbelief and a sense of absurdity, followed by rage and defiance, then worry, then the determination to make a lifestyle change, and finally acceptance.

A similar progression is occurring within our entire society. Our initial reaction to the terrorist attacks on the Twin Towers and the Pentagon was one of horrified disbelief and unreality— the sensation of entering a nightmare, as President Bush commented on his visit to New York in the immediate aftermath of the attacks—followed by rage, and then anxious worry. In response, many people at various times have contemplated some kind of lifestyle change. Here in Washington, dinner-party conversations always eventually drift onto the subject of whether it's time to relocate. After the Pentagon bombing, the anthrax scares, the sniper episodes, and the daily fluctuations in the threat levels of terrorist attacks, many people seriously contemplate moving elsewhere.

But before anybody calls in the moving vans, he or she should reflect on the experience of Wilmer McLean. In 1861, at the beginning of the Civil War, Wilmer became sufficiently anxious to flee his home in Manassas, Virginia, and move farther south to the comparative safety of Appomattox Court House. Four years later, on April 9, 1865, General Robert E. Lee and General

Ulysses S. Grant signed the final documents of Confederate surrender in Wilmer's front parlor. Sometimes—and this, too, is one of those times—there just isn't any refuge that offers guaranteed safety.

If flight isn't an option—and after we've successfully maneuvered beyond disbelief and feelings of absurdity and rage—then we're left with a limited number of options. Ponder for a moment that last quote by the retired engineer: "If you take precautions, you can live very, very well. You cannot just abandon that." Then answer the following multiple-choice question.

Do you believe the quote reflects:

1. An unrealistic and overly confident feeling that taking precautions—assuming one even knows what precautions to take—will prevent one from being killed or injured?
2. An expression of denial that evades any emotional response to possible pain, injury, or death?
3. A healthy expression of the humility we all should feel in response to the little control we all have in regard to what can happen to us from moment to moment?

The best answer to that question, it seems to me, is "all of the above." While it's true that no amount of precautions will ever guarantee safety, we can't allow our anxiety about an uncertain future to goad us into jettisoning our most cherished values and living in denial: shutting ourselves up in our homes or apartments without televisions, radios, or links to the Internet, and simply hoping for the best. What's demanded of us in the face of terrorism and other contemporary sources of anxiety is a major psychological realignment in our attitude toward anxiety itself. In a world where total security is impossible, we must learn to accept the fact that anxiety is going to remain a permanent part of our inner landscape.

"In an open society, there are simply too many threats, too many openings and too many interactions that are built on trust. You can't even begin to secure them all without also choking that open society. Which is why the right response, after a point, is not to demand more and more security—but to learn to live with more and more anxiety," wrote columnist Thomas Friedman.

Learning to live with increasing levels of anxiety means learning to take it in stride, becoming comfortable with the concept that feeling anxious is a "normal" part of living. Our attitude should be similar to that of the cancer survivor who learns to savor every moment despite the realization that however favorable the doctor's prognosis, the cancer may return.

What's needed is a new, more empowering approach that involves thinking in counterpoint. *The New Shorter Oxford Dictionary* defines musical counterpoint as "the art or practice of combining two or more musical parts in accordance with definite rules so that they are heard simultaneously as independent lines." Thinking in counterpoint involves simultaneously processing one stream of thought (our activities, goals, and concerns at the moment) while at the same time neither avoiding nor panicking at the secondary thought that we live in dangerous times in which total security is impossible.

As with the cancer survivor who thinks wistfully back to the time prior to diagnosis, our lives, too, have been irrevocably changed: We cannot return to the world that existed before September 11, 2001. The question is, are we going to build up our tolerance for anxiety—adjust our anxiostat, so to speak—and get on with our lives, or are we going to allow our anxiety to overwhelm us?

In *Poe's Heart and the Mountain Climber*, we will explore anxiety at every level from the molecular to the behavioral. Along the way, we will address such pivotal questions as these: How does anxiety differ from fear and stress? Which areas of the brain are associated with anxiety? Can animals become anxious? If so, how does animal anxiety differ from ours? And if we were free of all anxiety, would that be a good thing, or do we actually need a certain level of anxiety in order to be creative and live life to the fullest? Along the way, I'll provide some guidelines on how to manage anxiety, how to make it a positive rather than a negative influence on your life.

After interviewing many experts on anxiety, and reflecting on my own years of experience treating anxious patients, as well as experiencing more than a few anxious moments myself, I've organized this book around one principle: The best way to manage anxiety in these anxious times is to learn about it and put that learning to practical use.

While not intended primarily as a self-help book, *Poe's Heart and the Mountain Climber* contains suggestions I've received from experts on anxiety that, if you apply them, will enable you to manage anxiety in your own life. Thus, *Poe's Heart and the Mountain Climber* involves both information and its practical application. And although our exploration of anxiety will sometimes involve complex topics and principles,

I've tried to temper scientific precision with clarity and a user-friendly approach. As a first step, let's take a close look at two factors contributing to our anxiety: the deficient probability-estimating power of the human brain, and the increasing attempts by the media and others to influence our behavior by arousing our anxiety.

# 1

## Our Anxious Culture
*Triggers for Anxiety*

---

*U*NFORTUNATELY, OUR BRAIN isn't very proficient at probability estimation. Take an airplane phobia, for instance. Untold numbers of people suffer from a fear of flying, an anxiety condition that can range from the mildly discomfiting to the totally incapacitating. Most of us can bring to mind one or more acquaintances who refuse to step onto an airplane under any circumstances. More numerous are anxious flyers like myself who travel by air on a regular basis, but only in the absence of any reasonable or convenient alternative. Yet if you look into the statistics of air travel, behind all this you come up with a fairly astounding figure that logically should greatly reduce airline jitters.

Statistically, a specific air traveler would have to get on a commercial airplane daily for more than eight thousand years before falling victim to a multiple-fatality airplane crash. Death is much more likely to occur in the car used to travel to and from the airport. Car accident fatalities happen with a frequency of 1 in 18,800, with a significantly decreased risk if the traveler leaves the driving to a professional: Bus and train accident fatality statistics are 1 in 4,400,000 and 1 in 5,050,000, respectively. Motorcycles are associated with a 1 in 118,000 risk of death. Nor is walking the streets risk free (you have a 1 in 45,200 risk of being struck by a car).

While most of us experience some mild anxiety about travel risks, we tend to forget about the greater statistical risks involved if we confine our lives to the place where we feel the safest: our own homes. On lists of the world's most dangerous places, the home ranks second (the highway takes top honors).

In addition, we tend to be most anxious about grisly or horrific—albeit unlikely—possibilities. Think back a few summers ago when vacationers along East Coast beaches spent precious afternoon hours anxiously scanning the ocean waters for sharks. Death from a shark attack occurs at a rate of only about 1 in 94,900,000, a paltry number compared to death from drowning (1 in 225,000), skin cancer due to prolonged unprotected sun exposure (1 in 37,900), or even injuries from being struck by lightning (1 in 4,260,000). Despite these figures, many vacationers opted to play it safe by abandoning the beaches in favor of a few hours of boating—apparently oblivious of the fact that fatal boating accidents occur with a frequency of 1 in 402,000.

Even the most publicized of recent anxiety-provoking events involved more moderate risks than is commonly believed. While 2,801 people died in the World Trade Center attacks, about 15,000 people escaped the buildings; while 12 people died in 1995 after cultists released sarin nerve agent on three Tokyo subway lines, only 5,500 passengers out of the hundreds of thousands riding the trains that day required medical treatment; while 5 people died during the anthrax scare in the fall of 2001, infectious-disease experts estimate that many more people were exposed to the organism but failed to come down with the disease.

Here is a test (which I failed, incidentally) that can serve as a reality check on your own ability to accurately measure risk assessment. Please answer the following question about the likelihood of your becoming a victim of a terrorist attack: "If you won a free trip to one of the following places, which trip would you most likely pass up because of anxiety about personal safety: Israel, Istanbul, Bali, or New York City?"

Writer Wendy Perrin asked that question in late 2002 of 13,857 *Condé Nast Traveler* subscribers. Eighty-five percent felt Israel was too risky; 29 percent would avoid Istanbul, 56 percent wouldn't go to Bali because of the number of people killed there in the bomb blast earlier that year; but only 1 percent said they would pass on a free trip to New York.

While Israel scores highest in terms of perceived danger (only 15 percent of respondents would accept a no-obligation, no-strings-attached free trip) and New York seems quintessentially safe and universally desirable (99 percent of the respondents were ready to start packing their bags), statistics provide reasons for perhaps a more nuanced approach.

In the recent past, New York has seen more casualties from terrorism than anyplace in the world. In addition, most experts on terrorism place New York at the top of any list of potential terrorist attacks (Washington, D.C., comes in second, which, as a Washingtonian, I don't personally find reassuring). But despite the far more numerous casualties that accompanied the collapse of the Twin Towers, and the heightened risk of more attacks in New York in the future, almost all of Wendy Perrin's respondents said they would accept a free trip to New York.

Despite the meager 15 percent acceptance rate for a free trip to Israel, an argument can be made that even that conflicted and fragmented country is safer than New York—at least it was in 2002, when 202 Israelis had died at the hands of terrorists, compared to ten times that number of deaths in New York the previous year. Indeed, Israel's terrorism death toll— measured in fatalities per hundred thousand residents—is much lower than the annual homicide rate in New York and dozens of other U.S. cities.

What can explain these poll results? I suspect most people find New York less threatening than anywhere in Israel, Indonesia, or Turkey simply because New York is more familiar. As a rule, we tend to be most anxious when dealing with the novel or the unknown. This was true even before the emergence of terrorism; upcoming trips to new places creates in most people a mix of pleasurable anticipation coupled with a dollop of anxiety about whether events would proceed without a hitch.

In our more threatening world, it's only natural for us to envision something bad happening in unfamiliar rather than familiar places (one of the reasons videos of the fall of the Twin Towers still seem so nightmarish). The same thing holds true in regard to illnesses. We fear smallpox, anthrax, and SARS not only because they are so deadly, but also because we have no experience with such diseases. And the anxiety resulting from such uncertainty skews our thinking toward illogical conclusions.

Two factors determine the risks we're willing to take. The first is our risk perception—our estimation of the likelihood of a bad outcome. Access to information can influence risk perception, but only up to a point. For example, a year or so ago each of us was asked to decide this question: Should I take the smallpox vaccine? Experts on vaccines and public health weren't much help in deciding this question because of their disagreement about whether the remote threat of terrorist-initiated smallpox epidemic justified a mass inoculation.

As a result of this lack of expert agreement, each of us was left to decide on our own whether to find and take the vaccine. Most people didn't have a clue what to do and, as a result, felt anxious because they were being asked to make a potentially life or death decision about something they were professionally unqualified to evaluate or in many instances even understand. As a result, few Americans took the vaccine. Yet based on historical experience, such unwillingness doesn't make a lot of sense: The fatality rate from the vaccine can be expected to be no more than 1 in 750,000, a number that would be dwarfed by the fatalities and disabilities that would result from a terrorist-created smallpox epidemic.

Perhaps you consider a 1 in 750,000 chance of death unacceptable? If so, you should stop riding bicycles (1 in 341,000 chance of death) and stay out of swimming pools (1 in 225,000).

Risk tolerance, the second factor, is our willingness to accept foreseeable risks and move on. While two people may share a common risk perception (on many occasions a distorted one, as discussed above), one person may be willing to accept the risk, but the other won't.

For instance, several years ago I had an epileptic patient under my care who experienced frequent and life-threatening seizures despite trials on numerous anticonvulsants. Finally, I found a drug that worked. My patient's seizures stopped and, as a pleasing side effect of the new drug, she lost thirty pounds of excess weight. Everything was going swimmingly until six months after I started her on the drug.

While sorting through my mail one morning, I opened a letter from a pharmaceutical company sent to every neurologist in the nation warning of potentially fatal bone marrow and liver toxicity resulting from the use of my patient's drug. According to the letter, such complications were rare. Though troubling, the risk seemed acceptable to me; that is, if I were the patient, I would have continued with the drug. For one thing, regular checks on blood and liver function could detect these problems at an early stage. If the drug was then withdrawn, an affected patient would have a good chance for a complete recovery. But when I sat down with my patient and explained the situation to her, she opted to discontinue the drug because she considered the risks unacceptable. At her insistence, I switch her to another anticonvulsant. Over the next year, she experienced several seizures and regained the weight she'd lost.

What would you have done if you were in her situation? Would you have elected to remain seizure free and at an optimum weight for your age and height, but at a small risk for developing a serious side effect from the medication? Or would you have decided that the risk was simply too high? What would have been your risk tolerance in this situation?

In an attempt to gauge risk tolerance, Wendy Perrin asked the *Condé Nast* respondents a variation on her initial question: "How high a risk of a terrorist attack would you be willing to accept before canceling a pleasure trip?" More than half of them (53 percent) said they would not go if the odds were 1 in 100,000 or greater. Acting on the basis of such a low risk tolerance would preclude driving, walking the streets, or holding a job.

No, the human brain, it turns out, isn't very good at calculating odds. What's worse, it tends to overestimate the likelihood of rare, albeit painful or dangerous, experiences. And while this risk intolerance can provide a certain measure of security ("better safe than sorry"), it also deprives the anxious person of a great many of life's pleasures. Attempts to guarantee an unrealistic sense of security in an insecure world lead to many of the anxiety-driven behaviors we will discuss in more detail later in this book: phobias, obsessions, compulsions, panic attacks, and, most frequent, generalized anxiety disorder.

BUT THE MEDIA IS PROBABLY the greatest contributor to anxiety in our culture. In his book *The Culture of Fear*, sociologist Barry Glassner observed that "any analysis of the culture of fear that ignored the news media would be patently incomplete, and of the several institutions most culpable for creating and sustaining fears the news media are arguably first among equals." And while Glassner wrote about "fear," his observations are actually more descriptive of anxiety.

Certainly, on our local and national news programs, crime and disaster stories make up by far the greatest portion of the broadcasts. Nor are the newsmagazines any different. They adhere to the operating principle that we live in the midst of untold numbers of threats to our physical and mental health. As Glassner put it, "the guiding principle seems to be that no danger is too small to magnify into a national nightmare."

Television writers and commentators, it seems, have unwittingly adapted as their operating principle some advice once offered by Richard Nixon: "People react to fear, not love. They don't teach that in Sunday school, but it's true."

We have only to look around us to encounter applications of that philosophy: According to marketers, no home is safe without an elaborate burglar-alarm system; and increasingly sophisticated firewalls must be installed to keep out computer viruses and worms. While anxiety works as a great motivator, how much is too much?

Consider this question, asked in an editorial in the magazine *Anxiety Culture*, which advocates the maverick but nonetheless intriguing view that the anxiety level of the general population relates to the volume of sales of consumer goods: "Is there an optimum level of consumer anxiety (optimum from the point of view of the seller) at which the amount of sales of consumer goods is greatest?"

This "optimum level" would obviously land somewhere between two extremes. At one end of the scale, the consumer never buys anything because of being too scared to go outside or being too anxious to spend more money than absolutely necessary. At the other end of the scale, the person is so content already (i.e., he or she has a complete lack of anxiety) that there is no need to consume and no desire to purchase status symbols.

The editorial goes on to suggest that this optimum level of anxiety might be maintained by our constant exposure to the media's focus on crime and violence in news, current affairs, television drama, and films.

Nor is it just insurance and burglar-alarm sales that benefit from public anxiety. Some of us are willing to run up huge levels of debt to acquire those items we are led to believe might quell our social-comparison anxieties. Exactly *how much* we will go into debt before our financial anxiety overrides our social-status anxiety varies from person to person.

In short, in order to sell newspapers, boost TV ratings, or peddle commercial products, it helps to create anxiety in readers, viewers, and purchasers. Causes for that anxiety can range from comparatively trivial concerns such as whether you suffer from bad breath (and therefore should be popping "oral care" strips of Cool Mint Listerine throughout the day) to more weighty worries like whether this might be the day terrorists have chosen to carry out a deadly sarin attack during your daily subway ride to work.

And while anxiously mulling over these matters, you're likely to encounter on the morning news an advertisement encouraging you to undergo expensive laser surgery for the "correction" of the perfectly harmless condition of myopia (nearsightedness). Moments later you may encounter another advertisement suggesting that you can vastly enhance your appearance and sex appeal by buying a new pair of glasses. The hook here consists of the subtle suggestion that somehow your glasses are unbecoming. And although each of these two advertisements advocates an opposite course of action, they both share a common approach: arousing your anxiety that your present situation isn't acceptable, that you must either undergo corrective surgery or buy new glasses.

Health is another area of media-created anxiety. A *New York Times* article on the wisdom of maintaining stricter blood pressure control includes an interview with the coordinator of a national study on hypertension about his views on the benefits of exercise. Instead of providing factual information detailing the benefits of regular exercise, the reporter selects a quote guaranteed to arouse his or her reader's anxiety: "If you don't have time for physical activity, you will have time for illness. Illness doesn't make an appointment."

Reader anxiety is compounded a few sentences later by a second quote from that same expert that contains an admission that the suggested degree of blood pressure control "is simply not a realistic goal for many patients." Thus, the overall message is paradoxical and anxiety arousing: If you want to avoid high blood pressure—and the accompanying increase in the risks of heart attack and stroke—then you better make certain lifestyle changes. But even if you do so, you may be among the "many" for whom such measures aren't likely to make any difference.

Similar anxiety-arousing messages occur daily in newspapers and magazines in regard to cancer, Alzheimer's disease ("Are you frequently forgetting where you parked your car?"), heart attacks, strokes, and osteoporosis. The formula is familiar by now: Describe or illustrate an extreme example of an incapacitating, sometimes fatal, illness, suggest steps that you can take to avoid it, and end with the caveat that you may fall victim to the illness irrespective of what you do. For example, the SARS epidemic that felled large numbers of people in China in the spring of 2003 quickly spread to

Canada, raising anxiety that the disease might cross the border into the United States. Anxiety was further fueled by the fact that prior to the initial appearance of the illness, there would be no way of detecting it or controlling its spread into the general population.

Politicians, too, have learned the value of playing the fear or anxiety card. The basic theme was first set out in the advice given in 1947 by Senator Arthur Vandenberg to President Harry Truman on how best to garner congressional support for foreign aid to fight communist insurgents in Greece and Turkey: "Mr. President, the only way you are ever going to get this is to make a speech and scare the hell out of the country."

Truman took the advice, as did many succeeding presidents. Indeed, from 1947 through the fall of the Soviet Union in 1991 to the present day, the creation and maintenance of anxiety remains a regular part of foreign and domestic policy. At various times, the anxiety-inducing threats included nuclear war, race riots, gangs, and illegal drugs. Now we also have AIDS, herpes, hepatitis C, and other infectious diseases and, most notably, terrorism to be anxious about.

"Fear has once again become a powerful tool and motivator," wrote E. J. Dionne Jr. in a *Washington Post* article entitled "Inevitably, the Politics of Terror." But we don't just fear terrorism, we're anxious about it: "Will suicide bombings begin here in the United States? If so, what is the likelihood that I could be a target?" "Is airport security effective enough to eliminate any possibility that my airplane might be hijacked?"

Certainly, anxiety provocation is now an important component of our political process, a tool used by politicians whatever their persuasions. Some of our political leaders speak to us

repeatedly about impending terrorist threats ranging from suicide bombers to the employment of rapidly acting poisons and
deadly organisms. And our susceptibility to anxiety can be
used for both good (preparing people for an imminent terrorist
attack) and ill (playing on people's anxiety for strictly political
reasons). People *do* react to fear, and many of our politicians
are only too ready to play on that fear.

For example, President Bush's public remarks prior to the
beginning of the war in Iraq aroused general anxiety in our
nation that Saddam Hussein possessed weapons of mass
destruction (WMD). Bush and others insisted that not only
did Hussein possess such weapons, but that he was in a position to provide the weapons to terrorists. The general anxiety
aroused by this "imminent threat" to our security played no
small part in the decision to start the war. But despite intense
postwar efforts aimed at locating WMD, none were found—
certainly not on the grand scale that was confidently predicted
and cited as justification for attacking Iraq.

In response to this failure to find WMD, some commentators raised a question almost as anxiety arousing as the existence of WMD: Did President Bush and his advisers gain
acceptance for their plan to start the war by playing on our
anxiety about falling victim to terrorist attacks involving the
use of nuclear or other weapons? This question will, for a long
time to come, exert a polarizing effect on the nation. Without
taking a side in this debate, I think it's fair to predict that
given the intensity of the response to the WMD issue, our
leaders will hesitate in the future to advance personal political
agendas by playing the anxiety card. Nobody likes to be made
anxious.

## SUGGESTIONS FOR TRANSFORMING YOUR ANXIETY

I've ended certain chapters with some practical applications of the information contained in the chapter. I've also written more general suggestions that you'll find in the epilogue of this book. Some readers may elect to read these suggestions when finishing each chapter, while others may prefer to read all of the chapters and complete the book before looking at the suggestions.

### *Accept the Fact That We Truly Live in an Age of Anxiety*

Weapons of mass destruction, the prospect of nuclear Armageddon breaking out at flashpoints throughout the world, the war in Iraq and its consequences, the ever-present threat of terrorism, the rapid escalation of anti-Americanism throughout the world, lethal diseases spread by both natural means and as tools of biological warfare—these are only some of the potential threats to our national security, health, and economic well-being.

In addition, while we're deluged with advertisements warning us of the need for video cameras and other measures designed to enhance our sense of personal security, such persistent reminders of threatening scenarios do little to increase our sense of safety. Instead, they create additional anxiety: "Is our video camera sufficiently current?" As a result, anxiety has become a driving force shaping cultural, commercial, political, and health concerns, as well as our sense of international, national, and personal security.

At the moment, the anxiety created by various "threats" to our "security" is far out of proportion to the statistical likelihood that any one of us is going to fall victim to a terrorist bomb, a biological warfare agent, or a cancer traceable to those occasions when we've been in close proximity to a smoker. Despite the statistical unlikelihood of any of these things happening to us, our daily lives are filled with reminders of the possibility—however unlikely—that something awful (i.e., anxiety arousing) could happen. This free-floating form of anxiety creates a craving for information about threatening events that in some cases approaches the strength of an addiction. The 24/7 coverage of the terrorist attacks of September 11, 2001, now serves as a model: An anxiety-arousing event happens somewhere in the world; reports concerning it quickly escalate from news flashes to around-the-clock coverage to commentary that almost always puts the worst possible interpretation on the event. What's called for is the ability to learn new and important information while not allowing that information to induce overpowering anxiety.

*In Times of Crisis, Stay Informed, Since Your Own Imagination Can Create Images Even More Anxiety Arousing Than What's Really Happening*

Stick to news outlets that provide maximal information with minimal commentary and speculation. Your challenge is to keep abreast of world events so that you can make intelligent judgments concerning how those events may affect you and your family. Aim for balance. This has become even harder over the last few years now that many news outlets no longer make clear distinctions between "hard-news" reporting and commentary. As a corollary:

## Resist the Tendency to Become Obsessed
## with Disturbing News Events

After you've learned the basic facts, move on to other things, lest you become overly anxious, overwhelmed, or desensitized. That isn't always easy since we live in an age of information epidemics, or infodemics, according to intelligence analyst David Rothkopf. Like an epidemic, an infodemic results in widespread dissemination; but with an infodemic, the "organism" is misinformation—isolated facts mixed with speculation and rumor anxiously relayed worldwide via the instant communication made possible by the Internet, laptops, wireless phones, pagers, faxes, and e-mail.

Recent examples of infodemics include the SARS "threat," the frequent—almost daily—Department of Homeland Security reappraisals of the likelihood of additional terrorist attacks, and the reported existence of biological weapons stored in allegedly hostile Middle Eastern countries. In each of these examples, our anxiety transforms a legitimate concern into an imminent threat. It's only too easy to ignore the fact that few people died of SARS worldwide; that terrorist acts haven't become part of our daily life here in the United States; and that inspectors failed to find the weapons of mass destruction that served as the basis for going to war against Iraq. No doubt we can expect more, not fewer, infodemics in our future.

"This phenomenon is only going to grow more complex," says Rothkopf. "In the information age, life has changed fundamentally. Increased volatility is routine; events and information about them unfold rapidly; their consequences are amplified. The results are much like a roller-coaster ride: exciting, scary, disorienting and all rather different from the view from more solid ground."

Learn to view potential disturbing events from the ground rather than from the roller coaster. Personally, I concentrate on summaries of world, national, and local events while scanning the morning newspapers over breakfast. I skip the commentaries and editorials. On my way to work, I listen either to an all-music station or an audiotaped version of a current book. When at home at the end of the day, I rarely watch television; when I do, I avoid the talking heads who offer nothing more than guesstimates about the "meaning" of current news stories. In this way, over the years, I've saved myself untold hours of anxious fretting that I've been able to invest in more productive ways.

### Avoid News and TV Programs Aimed at Arousing Your Anxiety About Subjects You Can Do Little About

As noted earlier, the entertainment industry, which now includes news and information programs, recognizes the commercial value of anxiety. "Anxious people are inclined to eat and drink compulsively, need more distractions (newspapers, TV, etc.) and more propping up of their fragile self-image through 'lifestyle' products and status symbols," according to a commentary in *Anxiety Culture*.

As an example of how the media hypes viewer anxiety, recall the approaches taken by the major networks to the anthrax threat, the sniper crisis, and the war with Iraq. The coverage (especially on CNN) featured stark headlines incorporating words like *showdown* or *crisis*, accompanied by graphic videos of hospital isolation rooms, menacing-looking white vans, graphic war footage, and other anxiety-arousing scenes—all accompanied by background music marked by a mounting sense of tension and threat. "You, too, can fall victim to anthrax or bioterrorism, or the sniper" was the unspoken subtext. The best approach to take against this media assault on your equanimity? If you feel yourself becoming anxious while watching a particular show, turn it off. If you find yourself emoting rather than thinking about the news, stay away from the TV and seek news sources that provide you with information rather than subject you to an emotional roller-coaster ride. This is not a counsel for avoiding the news: Ignorance is not bliss. But your goal is to learn more, not to become more anxious.

# 2

# Mr. Snagsby Meets
# the Grand Inquisitor
*Anxiety versus Fear*

WHILE ANXIETY AND FEAR are often spoken about interchangeably, they differ in important ways. It was Charles Darwin who provided the first scientific explanation of fear. In *The Expression of the Emotions in Man and Animals* (1872), Darwin suggested that fear is an adaptive characteristic that evolved over millions of years and countless generations. Noting that fear varies in intensity from one occasion to another, Darwin suggested a fear continuum ranging from mild apprehension or surprise at one end to an extreme "agony of terror" at the other. Further, fear could be identified in animals and humans alike by such variable manifestations as trembling, dilated pupils, sweating, vocal changes, and peculiar facial expressions.

A quarter of a century later, Sigmund Freud made an important distinction between fear and anxiety, selecting the latter as the centerpiece of his theory of the psychoneurotic personality. According to Freud, danger could arise either from the external world or from one's own internal impulses. If the danger threatened from without—perhaps from an approaching hurricane or other natural disaster—the result was "objective anxiety," which corresponded to fear. In most instances, the emotional reaction was proportional to the potential danger: Larger risks engendered greater degrees of fear. But if the danger originated from within one's own mind, Freud called it "neurotic anxiety," which corresponds to the general meaning of the term.

In other words, while fear involved a specific, potentially harmful threat, usually a physical one, anxiety involved inner conflict. Freud described anxiety as "something felt," a state of chronic apprehension, "all that is covered by the word 'nervousness,'" such as feelings of tension or dread.

Initially, Freud settled on repressed sexual tensions as an explanation for anxiety. When blocked from normal expression, these tensions emerged in the form of free-floating anxiety. Eventually, he dropped this insistence on a sexual origin for anxiety and substituted a more general explanation based on anxiety as a danger signal. According to this explanation, anxiety acts as a motivation either to cope or to escape in response to the *perception* of danger in situations where there isn't objectively anything to be anxious about.

As an illustration of the fear-anxiety distinction, imagine yourself walking to an endodontist's office, where you will undergo a root canal. On the way, you encounter a supervisor from work, who passes you by without speaking. You're momentarily flummoxed by the apparent snub, but you forget all about it moments later when you're sitting in the dental

chair warily watching the oral surgeon prepare his instruments. At that moment, your fear of the dental procedure is at
the forefront of your consciousness; you're not thinking at all
about what happened on the street a few minutes earlier.

But that evening, if the operation went well and you're not
in too much pain, you're apt to find yourself running various
anxious scenarios through your mind in search of an explanation for the "snub." Why would the supervisor ignore you?
Should you start looking for another job? Or perhaps the
supervisor failed to recognize you because he or she was in a
hurry and mentally preoccupied—no snub had actually been
intended. Later, you may toss and turn in bed about the
"meaning" of what happened and how you should respond the
next morning when you encounter the supervisor at work.
You're not fearful about what might happen, as you were
moments prior to the dental operation; rather, you're anxious
about what you may experience in the morning.

Existentialist philosophers have pointed out another aspect
of anxiety that makes it different from fear. Anxiety can so
completely capture our attention that when we're in the
throes of an anxiety attack, we can't imagine a future moment
when we will feel differently. "Anxiety blots out time, dulls the
memory of the past, and erases the future," according to Rollo
May, psychoanalyst and author of *The Meaning of Anxiety*.
"While we are subject to anxiety, we are unable to conceive in
imagination what existence would be like 'outside' the anxiety." In this quote, May touches on the most painful aspect of
anxiety: the inability of the anxious person to imagine a future
moment when the anxiety will cease.

And there are other important distinctions. When we're afraid, we want to get away from the dreaded object as quickly as possible, whereas anxiety creates in us an inner conflict involving both approach and avoidance. When we're anxious, we keep turning over in our mind various "reasons" to explain our tortured inner state. But try as we might, we usually can't settle on any one reason for the persistence of this uncomfortable feeling that gnaws away at our peace of mind.

"Anxiety is usually linked with a strong *feeling of restlessness*," wrote psychiatrist-philosopher Karl Jaspers almost a century ago in his now classic 1913 textbook, *General Psychopathology.* "In a mild degree the state may occur in the form of a feeling that one has to do something or that one has not finished something; or it may be a feeling that one has to look for something or that one has to come into the clear about something." This emphasis on "restlessness" foreshadows the contemporary description of anxiety as currently set forth in the *Diagnostic and Statistical Manual of Mental Disorders*, fourth edition (DSM-IV) of the American Psychiatric Association: "restlessness or feeling keyed up or on edge."

Another difference: Fear also usually involves a short time frame—the fearful person either reaches the end of the dark alley unharmed or is mugged. Anxiety, in contrast, is prolonged and isn't improved by attempts at modifying the situation. As an example, consider a person who is always anxious around his or her boss. Attempts to avoid the boss only result in more anxiety, as do efforts to seek closer contact with the boss. No matter how benignly the boss acts, that anxious feeling doesn't go away. That's because the problem isn't the boss, but the anxious attitude held by the employee toward the boss.

Rollo May's description of this kind of conflict neatly distinguished between fear and anxiety: "In fear, your attention is narrowed to the object, tension is mobilized for flight; you can flee from the object because it occupies a particular point spatially. In anxiety, on the other hand, your efforts to flee generally amount to frantic behavior because you do not experience the threat as coming from a particular place, hence you do not know where to flee."

But the most striking difference between fear and anxiety involves the anxious person's subjective experience. In *The Concept of Dread* (recently retitled *The Concept of Anxiety*), Danish philosopher Søren Kierkegaard employed the word *angst* to describe an anxiety that has no specific object. Angst is related to anguish, which comes from the Latin *angustus*, "narrow," which in turn comes from *angere*, "to choke." For the anxious person, nothing less than life itself seems to be at stake. Indeed, Kierkegaard proposed that angst resulted from the acute awareness that one's existence can be utterly destroyed. The philosopher eloquently described the exquisite painfulness of anxiety: "And no Grand Inquisitor has in readiness such terrible tortures as has anxiety, and no spy knows how to attack more artfully the man he suspects, choosing the instant when he is weakest, nor knows how to lay traps where he will be caught and ensnared, as anxiety knows how, and no sharp-witted judge knows how to interrogate, to examine the accused, as anxiety does, which never lets him escape, neither by diversion nor by noise, neither at work nor at play, neither by day nor by night."

Unfortunately, it is sometimes difficult to distinguish between fear and anxiety without slipping into oversimplifications. For instance, in the fall of 2002, a massive manhunt was under way in the Washington, D.C., area for the sniper (it later turned out to be *snipers*) who had killed eight people during the previous eleven days. At the time, people were acting and reacting in unaccustomed ways: staying in their houses, and avoiding gas stations located near trees or other places offering the opportunity for concealment (four of the shootings occurred at gas stations). One of my patients told me that she was hiring a dog walker because of her fear of leaving her apartment. An Alexandria woman told a *New York Times* reporter that she was canceling plans to take her two sons to a nearby amusement park. "We're just not doing it. I'm just not going to wrestle with disaster," she said.

Nor was I immune. My writing desk looks out on woods, and I usually keep the louvered shutters open so that I can periodically pause and refresh myself by looking out at the trees. But during the sniper episodes, I kept the shutters closed. So, was I—along with my Washington neighbors—experiencing fear or anxiety? How does one decide when realistic fear slips along the continuum into the range of moderately severe anxiety? Consider the following personal experience at the time.

One morning during the sniper threat, my youngest daughter called from London, where she had been living for the previous year. She told me she was moving back to Washington. On hearing this, I experienced an immediate string of visceral discomforts: My heart seemed to miss a beat, I became aware of my breathing, and I felt that I needed to take an extra breath. I felt the need to sit down rather than remain standing, almost as if my legs weren't quite secure under me; I wasn't exactly dizzy, but I felt as if I might lose balance.

An equally discomforting mental scenario accompanied these physical discomforts. In my imagination, I pictured my daughter arriving home only to become the sniper's next victim, or perhaps being shot myself in the weeks prior to her arrival. I started thinking of speculations I had read in the morning paper that the sniper was a "psychotic psychopath" or perhaps a member of al-Qaeda or some other terrorist organization. I wasn't certain what to do or how to respond to all of this. Should I tell Ann to stay in London? But London wasn't entirely safe either, I reminded myself. While snipers weren't a threat, terror groups were. One could be killed or maimed for life by a bomb as a result of boarding the wrong bus or entering the wrong tube stop.

Obviously, for these few moments, I had slipped "over the edge": realistic concern yielded to anxious preoccupation, along with the accompanying physical sensations characteristic of anxiety. My acute anxiety emerged out of a background of chronic anxiety, which had begun two weeks earlier when the snipers killed their first victim.

But after a few moments, the feelings disappeared as I made a rough mental computation of the odds against the catastrophes I had been fantasizing about. In those moments, I stepped back over the boundary separating anxiety from realistic concern. I recognized the need to avoid extreme reactions like telling my daughter that it might be too dangerous to come home.

The bottom line is that in some instances a clear-cut distinction can't be made between the experience of fear or anxiety. For example, to the extent that the sniper threat was a real one, we in the D.C. area were experiencing fear. But the chance of any one person being selected as the sniper's next target (approximately 1 in about 500,000) was far less than the chance of getting into a fatal car accident or dying of a heart attack. On average, five thousand people a year die of food poisoning. Even the sun posed a greater risk than the sniper, with malignant melanoma felling almost eight thousand people a year. Yet nobody at that time was giving much thought to car accidents, heart attacks, food poisoning, or sun exposure—only to dying from a sniper's bullet. Nor were many people reassured by statistics. They were only too aware that statistics aren't worth hooey if you happen to be the one outlined in the crosshairs of the sniper's gun. Thus, you could say that we in the D.C. area were *fearful* of being personally affected by an extremely unlikely event and at the same time *anxious* at the thought that the highly improbable could still conceivably happen. In short, we were simultaneously fearful *and* anxious.

We should also distinguish, it seems to me, between stress and anxiety. One can be stressed and yet not feel anxious. If I'm upset by the sudden death of a friend, it seems fair to say that I'm responding to a stress. But later that evening, if I'm tossing and turning in bed at the thought that I, too, will someday die, I'm experiencing anxiety. Stress relates to all of the unpleasant things that can happen to us (illness, injury, social rejection, loss of possessions, and so on). Anxiety is how we relate to and handle such stresses.

Further, the terms *anxiety* and *stress* evolved from different origins. The roots of anxiety involve "perceptions" and "interpretations," purely psychological concepts. Stress, in contrast, can be traced to engineering, physics, and medicine: If exposed to sufficient stress, the bridge will collapse; excessive jogging may result in "stress" fractures. Not only do stress and anxiety have different origins, but they also frequently work in direct opposition to each other. Indeed, additional stress may sometimes bring relief from anxiety.

According to Rollo May, "In times of stress persons have something definite on which to pin their inner turmoil, and they can thus focus on concrete pressures. When there is great stress there may be freedom from anxiety."

As with most internal experiences, anxiety is hard to describe. While we know how our own anxiety *feels*, it's difficult to convey a sense of that feeling to others. So, instead of describing our feelings, we often find it easier to talk about the situations that arouse our anxiety. For most of us, these situations include one or more of the following: conflicts and frustration; threat of physical or emotional harm; challenges to our self-esteem; and pressure to perform beyond our perceived capabilities.

Whatever its cause, anxiety is an unpleasant emotion that we tend to describe with words like *worry*, *dread*, *fear*, or *apprehension*. And while the feeling of anxiety is sufficiently powerful to capture our full attention, it rarely provides clear guidelines on how we should respond. Typically, when we're anxious, we scan our environment in a search for clues about the source of our uncomfortable feelings. Some environments can almost be guaranteed to arouse anxiety.

"Descending underground at the Wall Street subway station, awe gave way to anxiety," wrote Mark Taylor, professor of religion at Williams College, in an essay in the *Los Angeles Times* about his first moments after leaving Ground Zero several weeks after the terrorist attack. "Anxiety is a response to what is there but not there, everywhere but nowhere—precisely like the webs and networks of terror, which now hold us in their grip. Individually and collectively we sense the danger of things slipping out of control and are not sure how or where to respond. Waiting for the train in silence, anxiety settled as thick as the dust enveloping us."

Philosopher James Park provides another perspective: "Anxiety is being afraid when there is nothing to fear. We struggle with something in the dark, but we don't know what it is. In angst we confront the fundamental precariousness of existence; our being is disclosed as unspeakably fragile and tenuous. And when it bursts through the protective shell in which we try to encapsulate it, our anxious dread renders us helpless."

Psychologist Derek Russell Davis picks up on this theme of foreboding: "An anxious person is in suspense, waiting for information to clarify his situation. He is watchful and alert, often excessively alert and overreacting to noise or other stimuli. He may feel helpless in the face of a danger which, although felt to be imminent, cannot be identified or communicated. Hope and despair tend to alternate."

Probably the most artful depictions of anxiety can be found in the works of Edgar Allan Poe. In my favorite story, "The Tell-Tale Heart," the unnamed narrator describes himself in the very first sentence as afflicted with a foreboding anxiety: "True!—nervous—very, very dreadfully nervous." Terrified by the eye of an old man with whom he shares a house ("He had the eye of a vulture. . . . Whenever it fell upon me, my blood ran cold"), the narrator decides to end his torturous anxiety by killing his housemate as he lies asleep in his chamber. But the victim unexpectedly awakes and sits upright in bed, listening. In fear for his life, he emits a "groan of mortal terror" that reminds the narrator of his own anxiety: "I knew the sound well. Many a night, just at midnight, when all the world slept, it has welled up from my own bosom, deepening with its dreadful echo, the terrors that distracted me."

Reveling in the old man's mounting terror, the narrator waits silently in the darkness of the hallway. Suddenly, he hears what sounds like the beating of the old man's heart. In response, "a new anxiety seized me—the sound would be heard by a neighbor! The old man's hour had come!" At this point he leaps into the room and kills his victim. After dismembering the corpse, he hides it underneath planks from the bedroom flooring.

The tale ends with the narrator sitting in the bedroom talking with three policemen dispatched to the scene to investigate a shriek heard by a neighbor. In the midst of the conversation, the narrator imagines he hears once again the increasingly forceful sound of the beating heart. At the end of the story, Poe describes the spectrum of anxiety the narrator is suffering, beginning with uneasiness ("But ere long I felt myself getting pale and wished them gone") and culminating in an "agony" of "horror." Overwhelmed by nothing but his own anxiety ("any thing was better than this agony"), the narrator finally blurts out a confession.

Charles Dickens provides a vivid description of a less severe but equally pervasive anxiety in his portrait of Mr. Snagsby in *Bleak House.*

Mr. Snagsby cannot make out what it is that he has had to do with. Something is wrong, somewhere; but what something, what may come of it, to whom, when, and from which unthought-of and unheard-of quarter, is the puzzle of his life. . . . And it is the fearful peculiarity of this condition that, at any hour of his daily life, and at any opening of the shop-door, at any pull of the bell, at any entrance of a messenger, or any delivery of a letter, the secret may take air and fire, explode, and blow up. . . .

For which reason, whenever a man unknown comes into the shop (as many men unknown do) and says, "Is Mr. Snagsby in?" or words to that effect, Mr. Snagsby's heart knocks hard at his guilty breast. He undergoes so much from such inquiries, that when they are made by boys he revenges himself by flipping at their ears over the counter, and asking the young dogs what they mean by it, and why they can't speak out at once? More impracticable men and boys persist in walking into Mr. Snagsby's sleep, and terrifying him with unaccountable questions; so that often, when the cock at the little dairy in Cursitor Street breaks out in his usual absurd way about the morning, Mr. Snagsby finds himself in a crisis of nightmare, with his little woman shaking him, and saying "What's the matter with the man?"

From Dickens's description, it's obvious that Mr. Snagsby cannot name what so tortures him. Instead, he is afflicted with a brooding sense of unease, a sense that something is somehow terribly wrong, something that may end in catastrophe for him.

As with Mr. Snagsby, the anxious person suffers the consequences—indeed, the agony—of perceiving the immediate animate and inanimate world, as well as his or her personal circumstances, as pervaded by a nameless dread. This sense of threat and vulnerability sets into motion various scenarios that vary according to background, education, and basic personality.

Under conditions that other people find mildly stressful, the anxious person may, as with Mr. Snagsby, suddenly explode into an inexplicable outburst of irritability followed by aggressive verbal—and even on occasion physical—attacks. Although Dickens couldn't put a name to it, Mr. Snagsby is experiencing anxiety in its purest form: generalized anxiety disorder (GAD), a free-floating sense of foreboding and worry that makes him only too ready to lash out at whomever at a particular moment stirs the embers of his profound disquiet.

According to the DSM-IV-TR, GAD is associated with the following: (a) excessive anxiety and worry (apprehensive expectation), occurring more days than not for at least six months, about a number of events or activities; (b) difficulty controlling the worry; and (c) the anxiety and worry are associated with three (or more) of the following symptoms: (1) restlessness or feeling keyed up or on edge, (2) being easily fatigued, (3) difficulty concentrating or experiencing the mind going "blank," (4) irritability, (5) muscle tension, and (6) sleep disturbance (difficulty falling or staying asleep, or restless unsatisfying sleep).

Among the various components of anxiety, worry is the most distinctive. Two main aspects of anxious worry distinguish it from normal worry. First, anxious worry is uncontrollable. In one interview survey, 100 percent of people with

GAD said they could not stop worrying no matter what they did. Second, anxious worrying is always excessive, unrealistic, and involves an overestimate of potential threat.

According to Michael J. Dugas, a psychologist at Penn State University, "Intolerance of uncertainty appears to be the central process involved in high levels of worry. GAD sufferers are highly intolerant of uncertainty. For example, they have told me things such as 'I know there is only one chance in a million that my plane will crash, but I can't help worrying about it because it might just happen.' I use the metaphor of an 'allergy' to uncertainty (where a very small quantity of a 'substance' leads to a violent reaction) to help them conceptualize their relationship with uncertainty."

Think of anxiety as an attempt to control situations that by their nature cannot be controlled. For example, although an accident is always possible during any flight, most people are reassured when they consider the extremely low risk that a specific plane may crash at any given time. Not so reassured is the anxious person whose anxiety stems from his or her inability to control the situation *except by worrying about it*. While the anxious person recognizes intellectually that his or her own attitudes have no bearing on whether an airline disaster occurs, he or she worries nonetheless. "When I fly in an airplane, I worry that the plane will crash and if I stopped worrying about it, it probably would crash," one GAD sufferer told his psychiatrist.

But the anxious person doesn't just worry about plane crashes and other disasters. "The one area that consistently distinguishes GAD patients from others is excessive worry about minor matters," according to Laszlo Papp, director of the

Anxiety Disorders Research Program at Hillside Hospital in Glen Oaks, New York. "A negative answer to the question 'Do you worry excessively about minor matters?' effectively rules out the diagnosis of GAD." And while anxious worriers tend to worry about the same things as everybody else, they routinely anticipate disastrous outcomes, especially under conditions of uncertainty or potential threat.

As an example of Papp's point, take triskaidekaphobia: fear of the number 13. Among the superstitious, Friday the thirteenth is an especially unlucky day. And while most of us lend little credence to such a belief, and take no particular precautions on those Fridays that fall on the thirteenth of the month, a case can be made that anxiety about Friday the thirteenth can serve as a self-fulfilling prophecy. For instance, the risk of dying in a traffic accident is greater for women on those Fridays that fall on the thirteenth day of the month. At least that was the finding of a Finnish national study of traffic deaths that occurred on Fridays between 1971 and 1997. According to the study, the risk of death on Friday the thirteenth is 63 percent higher than it is on other Fridays. The researchers speculate that the increased risk of death for the women on Friday the thirteenth results from their greater anxiety on that day. Even though they may not claim to be superstitious, their anxious anticipation of unfortunate events leads to the distraction that sets them up for fatal accidents. (The researchers for some reason didn't include men in their investigation.)

"Patients with anxiety process threatening information differently from other people," according to psychiatrist Thomas E. Brouette. For instance, since the anxious person feels helpless and inadequate under conditions of stress, events

at such times seem to proceed in unpredictable and uncontrollable directions. In an effort to cope, the anxious person
shifts attention from the outside world to self-evaluation. But
this only leads to further arousal.

Eventually, the anxious person develops what psychologist
Aaron Beck refers to as "automatic cognition": recurrent
thoughts involving harm or danger, especially under conditions
of stress. Thus, the narrator in "The Tell-Tale Heart" believes
that he hears the beating of the old man's heart building up to
an unbearable crescendo, whereas he is actually experiencing
nothing other than his own mounting anxiety brought on by
the presence of the policemen. Mr. Snagsby experiences a less
severe but equally pervasive anxiety, almost as if he were wear-
ing a pair of colored glasses that influences the hues and colors
of everything encountered in his surroundings. Every customer
entering Mr. Snagsby's shop represents a potential threat. What
do they want? Who sent them? Whoever the customers may be,
and whatever their reasons for entering his shop, Mr. Snagsby
finds them menacing and their arrival a sign of ominous developments yet to come. In such a world, everything is uncertain
and potentially dangerous.

SEARCHING FOR SOME ADDITIONAL help in understanding what
life must be like for persons inhabiting a world marked by constant anxiety, I consulted psychiatrist Michael Stone about the
etymology of the word.

The Romans in Cicero's time used the word *anxietas*,
which indicated a lasting state of fearfulness. This con-
trasted with *angor*, which signified a momentary sense of

intense fear, akin to our concept of panic. Angor also means strangling—and derives from the word *ango:* to press something together. The idea of narrowness is another connotation, as in the Latin *angustia* (narrowness), the French *angoisse* (anguish—a more acute, paniclike state), and the German *Angst* (fear) and *eng* (narrow). The *angr* root in Indo-European languages also gave rise to our *anger* (akin to Old Norse *angra:* grief) and *angina* (a term also used in Roman times to signify a crushing sensation in the chest and the accompanying dread.)

Tightness, closeness, confinement, suffocation, a feeling of constriction and pressure so intense that it threatens breathing—these are the very terms and phrases frequently employed by anxious patients when attempting to put their discomfiture into words. The sense of suffocation is a particularly common description and may be a holdover from our earlier evolutionary history.

According to psychiatrist Donald Klein, anxious individuals may be unusually sensitive to rising levels of carbon dioxide ($CO_2$) in their blood. Normal breathing involves fluctuations in $CO_2$ levels: decreasing levels as we breathe outward, followed by gradually increasing levels that peak just prior to the next expiration. But anxiety and even panic can ensue if the level of $CO_2$ reaches a certain point. Test this for yourself by holding your breath for a few moments; you'll experience a slowly evolving sense of suffocation leading, if you persist long enough in your effort, to anxiety and a sense of impending panic.

I remember experiencing a particularly intense $CO_2$-related episode of anxiety several years ago while snorkeling. In order to get a closer look at a coral reef, I took a deep breath and slipped beneath the surface. But I misjudged my depth and remained submerged a bit too long. As a result, while surfacing the ascent seemed endless; toward the last few moments, I felt that I couldn't hold my breath much longer. Anxiety turned into panic until that glorious moment when I finally burst into the sunlight. Klein would describe my experience as an example of an adaptive "suffocation alarm." The increasing levels of $CO_2$ in my blood raised a survival alarm that focused my attention on getting to the surface as quickly as possible in order to take a long deep breath.

Klein and others speculate that with some anxious people the sensitivity of the $CO_2$ alarm system may be altered to the point that even normal levels of $CO_2$ in the circulation can trigger acute anxiety, a "false suffocation alarm," as Klein refers to it. This hypersensitivity to $CO_2$ levels often leads to anxiety when encountering possible, albeit unlikely, suffocation situations.

For example, think back to the last time you were in a crowded elevator, or snarled in traffic while in a tunnel. Although inconvenienced by the delay, you probably didn't get upset. But a person with $CO_2$ hypersensitivity would respond differently in such situations. He or she would experience a sensation of suffocation, even though breathing wasn't compromised in any way. Heightened "respiratory awareness syndrome" is the medical term for this condition. But whatever term is used to describe it, the process involves becoming consciously aware of the normally automatic and unconscious process of breathing.

If you want to experience respiratory awareness syndrome for yourself, look up from this book and direct all of your attention to your breathing. Keep your attention focused on the effort it takes to expand your chest with each breath. Seems like a lot of work, doesn't it? Do you feel as if you're getting enough air? Perhaps you need to take a slightly deeper breath? If you focus intently enough, you may convince yourself that your breathing isn't sufficiently forceful and, in response, you're likely to take an additional, deeper breath. This temporary modification of your breathing pattern resulted from the simple act of directing your attention to a process that ordinarily you don't pay the slightest attention to. Indeed, many of the distinguishing features of anxiety result from similar intrusions into awareness of normally nonconscious processes such as the rate and rhythm of the heart, the steadiness of one's hands, or one's general sense of balance and coordination—processes that we alter, usually for the worse, whenever we become consciously aware of them.

Anxiety can also be stimulated in some people by sodium lactate, a chemical solution that can be found on the shelves of any emergency room, where it's used for the intravenous infusion of medications. While most people react to such infusions without any side effects or complications, sodium lactate often sets off a panic attack among people prone to anxiety.

The explanation for the lactate-induced panic can be traced to an observation about exercise first made several decades ago. Psychiatrists in the mid-1960s, at the start of the physical fitness boom in this country, began encountering increasing numbers of patients who experienced intense anxiety attacks during or shortly after vigorous exercise. Two psychiatrists speculated that the anxiety resulted from the buildup of lactate in the blood that normally accompanies vigorous exercise. As a test of their speculation, they intravenously infused sodium lactate into a group of patients known to suffer from anxiety. After receiving the solution, some of the recipients experienced immediate panic attacks that resembled their naturally occurring episodes. In contrast, the infusions produced no effect on people who didn't suffer from moderate to severe anxiety.

According to current thinking, lactate-induced panic stimulates the locus coeruleus, a brain-stem nucleus that manufactures the neurotransmitter norepinephrine. Another locus coeruleus stimulant, inhaled 5 percent $CO_2$, can also lead, as already mentioned, to an outbreak of panic in individuals prone to anxiety. Incidentally, don't allow this abrupt intrusion of technical terms to discombobulate you. In chapter 3, we'll explain (with a minimum of technical mumbo jumbo) the biology of anxiety, especially the brain structures that mediate the anxious response. But first, an even more fundamental question needs to be addressed: Are we the only creatures who experience anxiety? And if not, how far down the food chain can anxiety be traced?

# 3

## Scampering Sea Hares and "Inhibited" Children
*The Origins of Anxiety*

I F YOU'RE NOT TOO PICKY about definitions, anxiety-like responses can be observed at the level of single-celled organisms. A bacterium placed in a water-and-sugar solution will move toward the sugar molecules with a corkscrewlike motion. If a chemical poison is then added to the solution, the bacterium stops and after a moment moves off again in another direction.

Watching the bacterium's performance under a microscope, it's easy to conclude that the tiny organism has evolved the capacity to distinguish harmful situations from positive ones and to respond with anxious trepidation to the former. Not that different, it appears, from your willingness to pet your own dog while keeping a respectful distance from a strange dog whose responses you can't predict. But, as it turns out, such an analogy is fanciful; the bacterium's response is far less sophisticated and based on a simple mechanical model.

Attached to the bacterium's outer cell membrane are five to seven whiplike structures (flagella) and up to thirty chemical receptors. When the receptors detect specific molecules within the surrounding liquid medium, they activate the flagella to rotate in a propeller-like motion. If the encountered molecules are beneficial, the receptors orchestrate the flagella to rotate in the same direction; harmful substances, in contrast, induce the receptors to program discordant alternately clockwise and counterclockwise movements that propel the bacterium in a new direction.

"A change in behavioral state has taken place in the bacterium that differs from anxiety primarily in the simplicity of the information and memory processes taking place and in the narrowness of the repertoire of responses available," according to Columbia University psychiatrist Myron Hofer. "We have a scaled-down highly simplified prototype for anxiety, which is capable of producing some of the behaviors in anxious humans."

Although Hofer has a point, similar claims can be made, it seems to me, about a burglar alarm: "When I go into my house and plug in the correct code, my house experiences a comfy feeling of familiarity; but when another person enters and fails to come up with the code, my house experiences 'anxiety' and in response lets out a cry of alarm." In short, one must be cautious when seeking explanations and not become too enamored of superficial similarities lest one invoke that most ancient of bugaboos: anthropomorphism, the ascription of human form, attributes, or personality to something impersonal.

Take the research on *Aplysia californicus*, commonly referred to as the sea hare, an organism with a pedigree at least 100 million years more recent than the bacterium. In nature, the sea hare scampers away when confronted with its natural enemy, the starfish. In his experiments, Nobel Prize–winning neuroscientist Eric Kandel used an electric shock to create the equivalent of a starfish attack. He delivered the electric shocks in close temporal association with shrimp juice, a chemical the sea hare ordinarily ignores. After a while, the sea hare reacted to the shrimp juice alone by adapting defensive postures (withdrawing its gills, releasing camouflaging chemicals) as if in anticipation of an imminent electric shock. But in the absence of the shrimp juice, the creature reverted to acting like a normal sea hare.

One could speculate that during Kandel's experiment, an ordinarily innocuous substance had been converted into a signal that elicited a response very similar to the anticipatory anxiety you or I might feel if some diabolical scientist had carried out the same experiment on us. But again, as with the bacterium, a healthy skepticism seems necessary on the subject of whether the responses in these simple organisms are at all similar to the experience of anxiety in our own species.

Some more enlightening clues about anxiety come from experiments carried out on a creature we know the most about next to our own species: the laboratory rat, the descendant of those first small mammals that split off from their reptilian ancestors 250 million years ago. Since we, too, are mammals, we share similar brain structures (differently proportioned) with the laboratory rat. As a result of these structural similarities (homologies, as biologists refer to them), both humans and rats also share some of the same behaviors (differently proportioned, in most cases, although most of us can probably bring exceptions to mind).

Before proceeding to describe some experiments involving rats that provide insights into the biology of anxiety, I want to take a short "break from the action" to make a point.

Since few people hold much affection for rats, they have traditionally served as the perfect research subject: They can be prodded and poked and forced to endure all sorts of nasty experiences without getting many people disturbed or concerned. But attitudes have changed in recent times; increasing numbers of us hold serious reservations about experimenting on *any* animals, even one as generally despised as the rat. To which I would respond: Brain research, as with scientific research in general, tends to be imperfect and, unfortunately, oftentimes messy. That doesn't mean research methods can't and perhaps shouldn't be changed. But for our purposes in *Poe's Heart and the Mountain Climber*, I'm suggesting that we explore what we have learned about human anxiety as a result of animal experiments (not just involving rats, as you will see). Let's leave for another day and another forum our discussion about animal experimentation. To make this easier, let's return to our narrative and start with some innocuous experiments.

For example, both human infants and rat pups react to separation from their mothers. While human infants cry, rat pups emit ultrasonic distress calls aimed at reestablishing contact. Separation anxiety seems an appropriate term here: Both rat pups and human infants are anxious when Mom isn't in sight and call out in response to that anxiety. Any accusations of anthropomorphism here can be countered by a reference to an intriguing finding: Drugs capable of lessening anxiety in human adults also inhibit the distress calls emitted by rat pups when the creatures are separated from their mothers. On the flip side, drugs known to create anxiety in adult humans greatly increase distress calls in the rat pups.

So, considering the responses observed in bacterium, sea hares, and rat pups, when does it seem appropriate to begin employing the term *anxiety?* In other words, at what point in evolutionary development does anxiety begin? I'm referring here to the anxiety that under the proper circumstances, we, too, may experience. That question isn't so easily answered, it turns out. Part of the problem stems from the subjective nature of anxiety. We recognize when we're anxious because we can *feel* it; and we infer it in other people based on our observations of them—ranging from what they say to how they appear. But subjective identification, or empathy if you prefer, becomes increasingly difficult the further we stray from our own species.

One useful approach to this dilemma involves exposing animals to experiences we would certainly find anxiety provoking, observing their behavior, and then breeding the animals according to their responses to stress.

In the 1960s, researchers at the Maudsley, a psychiatric hospital in London, selectively bred a strain of rats that seemed to respond anxiously when stressed. Dubbed, appropriately enough, the Maudsley reactive rat strain, the creatures become quite undone when separated from their mothers: emitting ultrasonic distress calls; preferring to remain in one place rather than walk about; and defecating when placed in an open exposed location (a sign of distress in rats).

As a first step in distinguishing high from low reactive strains, the researchers observe the animals with the intention of getting a fix on their normal behavior. The simplest monitoring system consists of nothing more exotic than a researcher with a notebook and a lot of time to spare. After watching the animals for varying periods of time, the researcher writes down items of interest. After repeated observations, he or she constructs a profile of normal rat activity, which can serve as a standard for comparison with subsequent observations of other rats under various experimental conditions.

Despite the method's simplistic appeal, it leaves a lot to be desired. Human observers eventually become bored, give in to fatigue, or experience lapses in attention ("look-away errors" as they're called by the experimental psychologists). In addition, an animal's behavior often involves changes (distance moved, speed of movement, alterations in body posture) that, for even the most attentive observer, are difficult to estimate with the unaided eye. Moreover, while some animal behaviors last only a second or two, others extend over many hours. To get around these limitations, researchers now use video-tracking techniques.

According to a brochure for a high-tech video-tracking system, "You can track animals against any background and under any lighting conditions to visualize social contact over hours or even days." Typically, such systems are now used to videotape the animals as they move toward or away from one another, run mazes, or explore objects. Later, the experimenter reviews the video recordings in real time as well as in accelerated and decelerated versions. If everything goes well, the tracking system provides unique and otherwise unobtainable insights into specific behaviors—or at least that's how things are supposed to turn out.

In order to get a feeling for tracking techniques, I met in front of a computer monitor with information technology specialist Shannon O'Malley and watched a rat swimming around in what psychologists refer to as the Morris water maze. This setup, named after psychologist Richard Morris, is essentially the equivalent of a specially designed swimming pool for rats. In the middle of the pool, concealed beneath the surface of murky water, lies a circular platform. Since the pool is too deep for the rat to touch bottom, any animal placed in the pool has to keep swimming until it either drowns or discovers the hidden platform where it can rest.

The initial swim path of the rat we were observing looked like one of those patterns we create on scrap paper when we're trying to coax a steady flow of ink from a reluctant pen. This rat was obviously a slow learner and flailed erratically around the pool for several minutes before, seemingly by accident, discovering the much-needed rest stop. But the rat's sense of relief was short-lived. After a few minutes, the researcher reached down, picked it up, placed it back in the water, and then measured how long it took for the creature to relocate the hidden platform. On this second trial, the rat headed directly toward the platform—a neat demonstration of the rat's spatial memory for the location of the lifesaving water platform.

Since I was interested in anxiety rather than memory, I asked Shannon, my guide in all this, how an anxious rat might respond to the test. Her answer provided me with my new word for the day: "Some strains of rats or mice when stressed exhibit thigmotaxis—that is, they 'hug' the walls instead of exploring. As a result, they never find the platform. Let me show you a more specific test for animal anxiety."

Shannon pulled up on the screen a video in which the experimenter had placed several animals in an open space in the middle of a cage. After a few moments, some of them retreated to a wall and remained there while their more intrepid counterparts continued to occupy themselves in the open. The researcher then administered an anxiolytic drug (one that decreases anxiety) to the wall-huggers and observed any changes in their behavior. Sure enough, after the infusion of the drug the animals rejoined their comrades in the open space.

At this point, the experimenter placed a colored bead in the middle of the field. Some of the rats then scurried back to the wall while others stayed put and nosed around a bit, their curiosity overcoming their anxiety. But it took only a shot of the anxiolytic drug to restore the confidence of some of the wallflowers, and they moved back out into the open to nose around for themselves.

From here, we turned our attention to a video display of a rat pup's vocal responses to separation from its mother. Upon suddenly finding itself abandoned, the pup let out one of those ultrasonic "distress vocalizations" or "isolation calls" that because of their high frequency can't be detected by human ears. But despite the fact that we can't hear them, these ultrasonic calls can be detected and displayed by special electronic sensors.

Over the next few minutes, I watched several experiments aimed at analyzing the rat pup vocalizations. As the animals began the distress calls, the experimenter administered various chemicals and observed the effects. If the chemical led to a decrease in distress calls, the experimenter labeled it for further investigation as a potential anxiolytic (tranquilizer). If the chemical increased the distress calls, the experimenter tagged it as a possible agent for increasing anxiety (an anxiogenic).

Researchers are ultimately interested, of course, in reducing rather than increasing anxiety since, at least in humans, not much of a legitimate market is ever likely to exist for anxiety-inducing agents (except among terrorist interrogators, perhaps). But an anxiogenic agent could prove useful in animal research: Its chemical structure could serve as a template for the design of a drug to counteract it. The hope is that a drug found effective in reducing distress calls in abandoned rat pups might prove equally effective in relieving the treatment of an anxious patient.

But while drugs effective in relieving distress in rats may provide insights into how to manage human anxiety, the value of such observations is still marred somewhat by that old bugaboo—anthropomorphism.

"The human observer is inclined to identify the movement of one rat moving away from another rat based on an interpretation of the rat's presumed 'intention,'" observed animal pharmacologist Berry Spruijt. This leads, on occasion, to a situation in which the human observer of the videotape labels a rat's behavior as exploratory while the computer program interprets the same videotaped behavior as avoidance.

But anthropomorphism aside, one can't fail to be impressed while watching these experiments on rats in captivity at the similarities between their behavior and our own. When threatened, the "anxious" rat avoids open spaces, prefers familiar places, stops in its tracks when encountering anything potentially threatening, trembles, and emits those ultrasonic distress calls. With the exception of the distress calls, our responses to anxiety aren't that much different. When we're anxious, we tend to prefer to remain in our homes, limit our social contacts, hesitate when confronting anything new, and engage in nervous mannerisms, including mild tremulousness.

As I watched the rat and mice videos, I fantasized about a behavioral scientist carrying out a similar video-tracking experiment at a typical cocktail party. Prior to the party, the scientist installs hidden cameras that record the movements of the guests from multiple vantage points. Later, while viewing the movement of the guests on a computer screen in an adjoining room, the scientist observes that some of the guests are gregarious from the moment they arrive. Upon entering the room, they scan the crowd, looking for familiar people. If they don't recognize anyone, they step forward and introduce themselves to anyone who looks interesting. After a few minutes of conversation, they move on. Over the space of the first fifteen or twenty minutes, they make the rounds, skillfully "working" the crowd.

Other guests, in contrast, appear uncomfortable from the moment they arrive. Their discomfort immediately increases if they fail to spot anyone familiar. If they have come alone, they may go over to their host or hostess, engage in a short conversation, and then wait for the first opportunity to slip out unnoticed. If they arrive with a companion, they spend their time talking with him or her rather than circulating to meet unfamiliar people. Thigmotaxis is the name of their game. Stand back; don't speak unless spoken to; favor the periphery rather than the center of the room; and avoid eye contact with unknown people.

After a few minutes, our hypothetical scientist notices on his or her tracking apparatus some interesting changes in interaction patterns. As the guests imbibe their first drink followed by the second, and perhaps even the third, the anxious partygoers move away from the wall, begin socializing, and appear for the first time to be actually having a good time. Within the brief interval required for them to knock back a few drinks, the now gregarious guests have taken to flitting around the room in patterns that are interesting to this behavioral scientist, especially if he or she is a social interaction researcher.

This information would no doubt be interesting to members of the food and beverage industry, who may welcome the opportunity to sit beside the scientist observing patterns of hors d'oeuvres and alcohol consumption. And interesting as well to an anxiety researcher looking for ways of fine-tuning the long-standing observation that alcohol lessens anxiety: How long does it take for alcohol to exert an antianxiety effect? How much is required? What anxious traits will be the first to disappear? Who might be most susceptible to the anxiety-relieving effects of alcohol?

While my cocktail party experiment is hypothetical, the technology for carrying it out is already available to anyone who can afford a small concealed video camera and a tracking system. And since each guest in the experiment would be uniquely identifiable on the video screen, his or her movements could be monitored in real time, or, alternatively, movements occurring over several hours could be compressed into sequences viewable on a computer screen in thirty seconds or even less. Further, the experimenter would be able to program the tracking system to highlight any aspect of guest interaction that may prove valuable.

But despite some indisputable similarities between a presumably anxious animal's behavior and our own, analogies between animals and humans can only be pressed so far. Certainly, we can't be certain that animals as low on the totem pole as rats and mice experience humanlike anxiety. So far no experimenter has come up with a rodent capable of talking to us about its inner world. It takes a vivid imagination on our part to compare a rat scurrying hastily about in its cage in response to a loud noise with, say, an executive pacing the length of his or her office while contemplating the prospect of being let go from his or her job.

ONE APPROACH TO SPANNING the conceptual gap between animal and human anxiety is to move up from rats and mice and concentrate on creatures that share with us many genetic and behavioral characteristics. Rhesus monkeys are a favorite choice of scientists since, besides sharing somewhere in the range of 90 to 99 percent of our genes, their brains and behavior are sometimes unsettlingly similar to our own. But there are, of course, limits: Rhesus monkeys living in cages differ in significant ways from their counterparts in the wild (omitting altogether, for the moment, the differences between caged monkeys and ourselves). As a result, if you want to study monkey anxiety as it occurs in real life, it's necessary to observe the monkeys in their natural state.

In order to do this, neuroscientist Stephen Suomi created a free-roaming colony on an island in the Caribbean. In the course of his breeding experiments, Suomi observed that even under these quasi-natural conditions, about 20 percent of the monkeys in the colony developed most of the signs of a mild anxiety disorder. As infants, they hovered near their mother and showed little curiosity about their surroundings. Later, as juveniles, they became undone whenever left by their mother to their own devices. At such times, they become agitated, then lethargic, then withdraw from other monkeys, and then, finally, assume a fetal-like posture. And while these anxious features are heritable, they can be profoundly influenced for good or bad by the animal's early experiences.

For instance, if raised by peers instead of their mothers, the anxious monkeys became so sufficiently withdrawn that upon reaching adulthood, they occupied the bottom of the hierarchical totem pole—an impressive demonstration that an unfavorable environment only worsens the harm inflicted by "bad" genes. Fortunately, most of this harm can be avoided by suitably altering the environment. When Suomi placed the anxiety-prone infant monkeys with unusually caring and experienced foster mothers, they grew up normally, and in some instances even rose to the top of the dominance hierarchy. Personally, I find these later findings tremendously hopeful: Genes are not destiny. Instead, life experiences can favorably modify the influence of genes on the subsequent development of adult anxiety.

When I first learned about Suomi's research, I reflected that a tendency toward anxiety also exists among our domesticated pets. A typical example of an anxious pet is my own African gray parrot, Toby. With the exception of the first few weeks after hatching, Toby has spent his entire eighteen years with us. And when I say "with us," that is exactly what I mean. Like all parrots, Toby is a flock animal. His inherited need for companionship leads him to become fretful and demanding when alone. If left even temporarily to his own devices, Toby turns restless, then calls out my name, followed successively by the names of my wife and three daughters. If these efforts at attracting someone prove unsuccessful, he resorts to a high-pitched, ear-and-nerve-jarring screech.

After observing my bird and his response to being left alone, I've concluded that Toby's intolerance for solitude qualifies as an anxiety response. For proof of this, one need only observe Toby's behavior when his entreaties successfully summon someone to his cage. From the moment Toby's rescuer responds to his entreaties, picks him up, and talks soothingly to him, the bird's demeanor changes: His eyes cease darting about, his ruffled feathers smooth into place, and he assumes a relaxed, easy posture. That body language conveys the unmistakable message that Toby's isolation-induced anxiety attack has ceased.

As it turns out, my bearded collie, Bobbie, also suffered as a puppy from isolation-induced anxiety. When I bought Bobbie, the breeder casually mentioned that the pup seemed "overly attached" to his mother. Whenever the mother walked out of the room during the first few weeks of Bobbie's life, he would follow her; the other pups from the litter, in contrast, just stayed put.

After I brought Bobbie home, I soon confirmed what the breeder had told me. The puppy dreaded being alone; when left to himself, he whimpered until brought to wherever in the house I happened to be. As Bobbie grew older, these signs of separation-induced distress ceased, only to be replaced by panic attacks during thunderstorms.

Bobbie is now eleven years old, and, while he no longer becomes quite as anxious when no one is nearby, he continues to come completely undone during storms. At the first sign of a storm, Bobbie paces about, trembles, pants, lets out a series of increasingly loud and distressed-sounding barks, tries to run from the house if inside, or, if outside, hurls himself against the door until someone lets him in. In fact, an actual thunderstorm is no longer required for him to develop pronounced symptoms of canine anxiety. All that's needed for him to anxiously seek solace either with me or another member of the family is an overcast sky or other indicators of an upcoming storm. And if we happen to be away from the house during a thunderstorm, we can be certain that upon our arrival we will find Bobbie cowering in the basement. Nor does he entirely respond to reassurances and petting at such moments. He continues to tremble and hyperventilate, sometimes requiring physical restraint lest (as happened on several occasions) he break away and fling himself against a screen door, tearing out the screen in the process. These canine panic attacks accompanied by flight occur with sufficient frequency that Bobbie has become a familiar figure to rescue workers at the local animal control facility.

Toby and Bobbie are not unique. Geneticists have identified in many breeds an "inbred anxiety" similar to Bobbie's. Dogs of different breeds, it turns out, respond with varying degrees of "timidity" (as the vets refer to it) when approached by a stranger or when encountering something new. Among basenjis, beagles, cocker spaniels, Shetland sheepdogs, and fox terriers, basenjis are the most timid; cocker spaniels, the least. Among Labradors, Australian kelpies, boxers, and German shepherds, the shepherds prove the most skittish when encountering novel situations, while the Labradors are the least fearful.

Cats, too, vary in their tendency to be fearful or aggressive. About 15 percent of ordinary housecats are standoffish around strange people and also fail to attack rats, while a larger group (about 40 percent) cozy up to strangers and have no hesitation in going after their rat enemies. Not only dogs, cats, parrots, and rats, but also wolves, cows, rhesus monkeys, and even paradise fish differ in "timidity" when encountering something unfamiliar.

Of course, any claim that Bobbie is experiencing anxiety flies in the face of the commonly expressed opinion that although animals can experience fear in the face of threat, they cannot experience anxiety since they aren't capable of mentally projecting themselves into the future. In a nutshell, the argument breaks down to this: If the animal isn't capable of imagining anything beyond the immediate present, how could it be anxious? And what could it be anxious about? Simple observation of Bobbie's behavior disproves this simplistic distinction. Bobbie starts behaving anxiously several hours before the arrival of a storm—in many cases, even before the first signs of bad weather.

Perhaps it's best to think of fear and anxiety in animals this way: For the most part, animals in the wild experience fear rather than anxiety. But with domestication (pets) or confinement (animals bred for experimentation), fear imperceptibly blends with anxiety. Have these animals become more anxious as a result of being deprived of their natural habitats? Or could this newfound susceptibility to anxiety be related to the fact that they share their lives with anxious creatures like ourselves? I don't pretend to have an answer to that question. There's no doubt, however, that we humans more often experience anxiety rather than fear. That's because while we rarely encounter direct threats to our physical well-being, we regularly imagine innumerable calamities that will never happen. Moreover, this tendency to catastrophize everything around us can be traced to our earliest years.

ALTHOUGH TO THE CASUAL OBSERVER one infant and young child may appear to behave very much like another, mothers and fathers can recognize vast differences in their children from the moment of birth. Most notably, children from their earliest years differ from one another in temperament: their moods and behavior. But since an infant can't speak of its feelings, and can express only a limited behavioral repertoire, judgments about temperament must be inferential. The most direct approach—simply observing a young child's behavior—reveals that the most striking temperamental difference is the child's characteristic response to unfamiliar people, objects, and situations. At the extremes are two categories of infants and children.

Jerome Kagan and his associates at Harvard University have identified a small group of shy, fearful, timid infants (estimated to be about 10 to 15 percent of the infant population) who grow into anxious adolescents and later into shy, cautious adults given to anxiety and even periodic panic attacks. These "inhibited children," as Professor Kagan refers to them, are distinguished from others on the basis of birth order (they are usually found among the younger siblings), physical appearance, behavior, and physiology. For this group, in general, arousal takes place more easily within those areas of the brain that process fear.

Inhibited children tend to respond to novelty with faster heart rates, increased muscle tension, and higher levels of some of the body's excitatory neurochemicals. "We believe that most of the children we refer to as inhibited belong to a qualitatively distinct category of infants who were born with a lower threshold for arousal to unexpected changes in the environment or novel events," says Kagan. "For these children, the prepared reaction to novelty that is characteristic of all children is exaggerated."

While Kagan's studies aren't the equivalent of a selective breeding experiment—ethically impermissible in any case—they do provide evidence that 10 to 15 percent of infants and children become subdued and often distressed when faced with novel situations or people. Do these "inhibited" infants and children grow up to produce children with similar dispositions? Extremely inhibited children, it turns out, are at an increased risk for developing anxiety disorders as adults. The picture is less clear when it comes to children on the mild-to-moderate range of the continuum.

And what about the genetics of all this? Is a tendency for anxiety genetically as opposed to environmentally inherited? In other words, does the anxious child experience greater-than-normal anxiety because he or she has inherited a gene or genes for increased anxiety? Such a question isn't so easily answered. Even a finding of increased anxiety among the children of anxious parents wouldn't be of much help in separating nature from nurture. While it's true that the "inhibited" responses of children could certainly result from genes inherited from one or both parents, those responses could also result from daily exposure to parental shyness, withdrawal, caution, and a generalized low-intensity anxiety.

Unfortunately (or fortunately, depending on how you look at it), it's never possible to cleanly separate genes from environment. To date, no one has discovered a set of genes that isn't operative in some sort of environment. Thus, it shouldn't come as a surprise to learn that genetic statistics on anxiety suggest both genetic and environmental influences—a blend of nature and nurture.

About 20 percent of first-degree relatives of people with generalized anxiety disorder (GAD) also have the disorder, compared with only 3.5 percent in the families of people not afflicted with the disorder. While such a difference strongly suggests a hereditary contribution, the environment must also play a role, since four out of five relatives are perfectly normal. (That figure is in line with Suomi's finding of a trait for hereditary anxiety in about 20 percent of the monkeys in his colony.) But predicting which one of five pups, rhesus monkeys, or humans will grow up to become an anxious adult remains an art rather than a science. (Recall that Bobbie's littermates seemed perfectly content when left on their own.)

Among twins, the incidence figures on anxiety are higher: If one member of the pair has GAD, the odds are 30 percent that the other member of the twin pair will also have it. But more is involved here, too, than genetics, since a clear-cut difference hasn't always been found between identical twins (sharing exactly the same genes) and nonidentical twins (sharing only some of the same genes). If anxiety followed a strictly genetic pattern, the incidence in identical twins should approach 100 percent, even among children raised in different adoptive families. This absence of any consistent difference in the incidence of GAD between identical and nonidentical twin pairs emphasizes the importance of environmental factors: Children become anxious adults as a result of dealing on a daily basis with an anxious parent or parents. Perhaps the fairest summary of the nature-nurture situation is this: Available evidence suggests a strong hereditary, if not necessarily genetic, component to anxiety.

Back to Kagan. According to his research, an inhibited temperament in early childhood increases the risk for the later development of one of the anxiety disorders. And since the inhibited child tends to avoid novelty and the uninhibited child seeks it out, it seems sensible to expect differences in brain activity—but what are those differences? And wouldn't it be nice to be able to come up with a test of brain activity that would make possible the early identification of a child as inhibited or uninhibited?

During a discussion that took place a decade or so ago, I asked Jerry Kagan to speculate about which brain area would turn out to be most important. He chose the amygdala. (We will have much to say about the amygdala in chapter 5, but for now it's sufficient to know that the amygdala [plural: amygdalae] is associated with the perception and expression of emotion.) In 2003, Kagan and his associates put his speculation about the importance of the amygdala to the test using functional magnetic resonance imaging (fMRI), which provides a near instant measure of ongoing brain activity.

They chose as research subjects twenty-two adults (mean age just shy of twenty-two years) who had been described at age two as inhibited (thirteen people) or uninhibited (nine people). In Kagan's experiment, the subjects looked at a series of pictures of faces. Some of the faces were familiar, while other faces had never been previously encountered.

The thirteen adults identified as inhibited at age two showed a significantly greater response in both the right and the left amygdalae to novel faces when compared to the nine adults who had been categorized as uninhibited. This difference didn't occur when all twenty-two people participating in the experiment looked at familiar faces; the activity in the amygdalae looked the same in the two groups. The findings provided an impressive confirmation of Kagan's hunch. Moreover, the implications of this research are quite profound.

"These findings support the hypothesis that inhibited and uninhibited infants are characterized by different amygdalar responses to novelty and suggest that some brain properties relating to temperament are preserved from infancy into early adulthood," wrote Kagan and his colleagues in their *Science* report.

We have here a foreshadowing in capsule form of one of the themes of this book: Not only does the brain of an anxious person differ from his or her calm cool and collected counterpart, but it's possible to correlate anxiety with the activation of specific brain structures and circuits. We know this on the basis of animal research (sorry about that) aimed at discovering the underlying brain activity associated with fear, anger, and aggression. We will describe this research in the next chapter, which opens with a personal story.

## SUGGESTION FOR TRANSFORMING YOUR ANXIETY

*As an Anxiety Antidote, Think of Anxiety as an Experience That Varies in Intensity from One Person to Another and from One Time to Another*

Acknowledge and accept that just as you were born with certain physical characteristics that either can't be modified at all or only by drastic measures, you may also have been endowed at conception with a genetic predisposition toward anxiety.

Your genes greatly contribute to how much anxiety you experience; if you're an overly anxious person, it's quite likely that your parents were overly anxious as well. This shouldn't lead to despair or feelings of helplessness, but, rather, acceptance. As we grow older, most of us learn to accept and work with inherited personal qualities. You should take a similar attitude toward your anxiety. While some people are fortunate and experience anxiety levels that are easily manageable and appropriate to life circumstances, others experience anxiety out of proportion to any reasonable assessment of the likelihood of potential harm.

Think of anxiety as a symptom similar to, say, a sore knee or a "bad" back. Since the knee or back pain may recur from time to time, you eventually learn not to give in to despair, but, instead, manage the symptoms by making minor adjustments to your daily routine. Take the same approach to anxiety.

Learn to use your own personal level of anxiety as a motivator: Schedule periodic "anxiety checks" throughout the day, during which you monitor your level of anxiety.

Since anxiety differs in intensity from one person to another, assess where you stand on the continuum extending from mild, infrequent episodes to intense, daily, and sometimes disabling anxiety. Perhaps, at some moments, anxiety may be difficult to bear and affects every aspect of your life. At other times, it may be barely perceptible, merely hovering in the background of your internal experience.

Of course, you can fret and fume about your placement on the anxiety continuum. "If only I could remain calm and in control like some of the other people around me," you can say to yourself—and then become additionally anxious that you can't create an anxiety-free persona for yourself. A more useful approach is to accept the fact that you are not personally responsible for the level and intensity of your anxiety.

# 4

# Startle Responses and Halloween Cats
*The Brain Circuitry of Fear and Anxiety*

WHEN I WAS ABOUT TWELVE years old, my younger sister played a trick on me that I still vividly remember more than forty years later. During an evening thunderstorm, she stepped out the front door, rang the doorbell, and then slipped back inside and hid herself in a small alcove. Moments later I came to the door and apprehensively peeked out the window into the darkness. Who could be ringing the doorbell on such a night? At this point, my sister leaped from behind me in the darkness and shouted, "Boo!" I was so frightened that my legs gave out from under me and I landed on my knees.

My reaction to the sudden fright didn't involve any conscious deliberation about the origin of the loud voice. I startled before I knew what had frightened me. Only after I had fallen to the floor did I finally put together what had happened. Thus, neither fear nor anxiety would seem to convey the essence of what had happened to me. I wasn't frightened of any specific threat (fear), since I hadn't the time to make any such formulation. At the same time, anxiety hadn't played a part, either, since I had been completely relaxed only moments before my sister's prank. The simplest explanation of what happened is that I was the unsuspecting victim of my own startle response: My body had reacted to the loud voice by falling to the ground as the result of the automatic activation of a self-defensive reflex.

In the 1930s, two psychologists, Carney Landis and William A. Hunt, made a formal study of the startle response. By today's scientific standards, their methods seem like something out of Laurel and Hardy or the Keystone Kops, or a bizarre variation on my sister's prank. One of them would sneak up behind an unsuspecting subject and fire a blank pistol while the other would film the subject's response. On the basis of this zany approach to research, they discovered a reproducible startle pattern that varied little from one person to another. It consists of a general flexion of the body, "which resembles a protective contraction or 'shrinking' of the individual," accompanied by eye blinking, forward head movement, a raising and drawing forward of the shoulders, a raising of the upper arms away from the body, and a bending of the knees.

Landis and Hunt insisted that the startle response was "pre-emotional . . . a rapid, transitory response much more simple in its organization and expression than the so-called 'emotions.'" In other words, emotions *follow* rather than initiate the startle response. Following their initial startle, the subjects in the Landis-Hunt experiments reacted with fear followed by annoyance. The startle sequence went specifically like this: an automated bodily response followed by fear, irritability, and, finally, anger. Certainly this sequence jibed with my own experience: My initial emotion after my bodily response of falling to my knees was a rush of anger directed at my sister. Although I didn't know it at the time, my fear-anger-aggression response mirrored animal research findings dating to the 1960s.

In one experiment worthy of a horror movie, scientists delivered shocks to two rats placed on an electrified grid until the animals began moving toward each other. The current would then be abruptly cut off so that each rat, presumably, would pleasantly associate the termination of the shock treatment with its close physical proximity to the other animal. But things didn't work out quite as the experimenters expected.

Instead of continuing to move closer together, the animals attacked each other before the shock could be turned off. And the greater the intensity, duration, and frequency of the shock, the more aggressively each animal attacked. Further, this aggressive response occurred across species lines. Mice, hamsters, monkeys, alligators, and even boa constrictors went at each other when electrically shocked. The same response occurred when different animals were caged together: a rat with a guinea pig, a monkey with a rat, and a rat with a rooster.

Nor did electric shock turn out to be the sole motivator of aggression. Withholding expected food rewards from the animals also provoked them to attack their neighbors (as well as any object at hand, such as stuffed dolls and tennis balls) with a ferocity matching their response to the foot shock. To the researchers' surprise, depriving animals of anticipated rewards worked as effectively as physical pain in provoking aggression.

(Surely, during their more reflective moments, the experimenters felt some discomfort themselves at so discomforting a fellow creature that hadn't done anything to them.)

We can observe human applications of this early animal research all around us. Newspaper stories appear with distressing regularity describing violent responses to the loss of "anticipated rewards" incurred when a person loses a job or is jilted by a romantic partner. In such cases, the pain is psychological rather than physical: fearfulness about what the future may hold, loss of face, diminished self-esteem, and so on. In our own species, it seems, psychological pain in response to ego affronts is every bit as motivating as any physical discomfort.

"Somehow we have developed in our minds a crucial linkage between even minimally measurable affronts to our status and the very fact of our survival," wrote Willard Gaylin in *Hatred: The Psychological Descent into Violence.* "We are more likely to feel threatened by an assault on our reputation, our status, our livelihood, our manhood—or even a misperceived assault in these areas." And, given the fact that in Gaylin's words, "we all live in the world of our own perceptions, where reality is only an occasional intruder," our anxious responses too often turn rageful and violent. Where in the brain does this rageful response originate? And why is it so intimately connected with fear and anxiety?

To answer these questions, it's necessary to describe one more rather grisly experiment (no more after this, I promise), this one involving cats. Surgical removal of the animal's cerebral cortex creates a kind of demon cat that responds with fear and anger to anything new and thus potentially threatening.

The "decorticate" (the technical term for the operation) cats crouch down, arch their backs, retract their ears, unsheathe their claws, growl, hiss, and attack any object in the vicinity. In addition, their autonomic nervous system goes into overdrive: pupils dilate, heart rate accelerates, blood pressure spikes, and hair stands on end. To get the picture, imagine yourself encountering the Halloween cat of your nightmares.

If cats respond with exaggerated rage after the loss of their cerebral cortex, then this might suggest—applying a little deductive logic here—that the cortex ordinarily reins in overly fearful and aggressive responses. Thus, the rage tantrums represent failures on the part of the cats to suppress their fears and the resultant rageful responses. Destroy the cat's cerebral cortex and you wind up with unmitigated fear leading to rage and then attack.

Reduced to its essentials—and sparing you, as promised, descriptions of untold numbers of search-and-destroy missions carried out on the brains of animals ranging from pigeons to Barbary apes—the operative principle goes like this: Beneath the cerebral cortex lies a network of brain structures (see Figure 1) that provide the underpinning for emotions. And while our *experience* of the various emotions arises spontaneously based on the activation of this network, our *expression* of these emotions depends very much on the contributions from the cerebral cortex. Thus, while we retain limited control over the experience of emotions, we remain largely in control when it comes to emotional expression. A moment's reflection on life's everyday experiences demonstrates the validity of this arrangement.

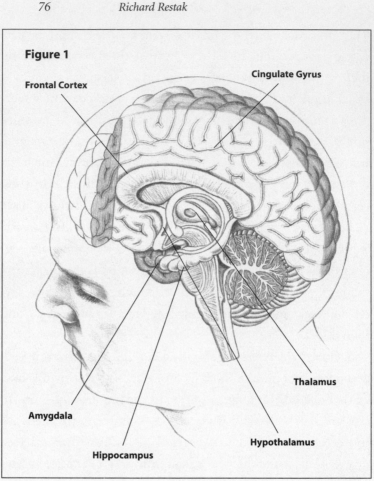

**Figure 1**

Frontal Cortex

Cingulate Gyrus

Thalamus

Amygdala

Hypothalamus

Hippocampus

If someone insults your spouse, you may insult the person back, or, more likely, you'll decide he or she isn't worthy of your attention and walk away. If you choose the latter, you do so because your cerebral cortex has exerted a restraining influence on the emotional network dwelling deeper down in your brain. But if you're overstressed or dreadfully insecure, you might express similar thoughts about his or her spouse in return. Your risk of engaging in such foolishness would be even greater if you've had too much to drink; when that happens, the alcohol disrupts the normal modulation usually exerted by your cerebral cortex. If the modulation is sufficiently impaired, you may lose emotional control altogether and physically strike out at the insult bearer.

Think of the cerebral cortex and the underlying emotional circuit as coexisting in a balance that given the proper circumstances can tilt either way. Too much cortical control and you'll experience and express little emotion; too little control and you're overreacting to everything that's happening around you.

Included in the emotional circuit are the thalamus, the hypothalamus, the hippocampus, and the amygdala. (See Figure 1.) Experiments on these structures involving either electrical stimulation or tissue destruction (ablation, as it's euphemistically referred to) reveal that all of them are associated in some way with emotions. The problem is that several of them serve other purposes as well. The hippocampus, for instance, provides the initial encoding for memories; the hypothalamus regulates, among other things, body temperature, blood pressure, and the blood levels of a host of chemicals, including sugar and hormones. How then does one tease out the contributions of each structure in order to develop a unified field theory of emotions?

STARTING TWO DECADES AGO, brain scientists in search of the neural basis of fear started using the conditioning procedures that had long been popular with behavioral psychologists. Conditioning started with the Russian experimental neuropsychologist Ivan Pavlov and his now classical experiments demonstrating that dogs can be made to salivate to a bell sounded just prior to feedings. Scores of experiments carried out since Pavlov's work with dogs have led to conditioning in animals ranging from starfish to humans. Like many scientific procedures, conditioning can be described using a lot of mumbo jumbo, but the principle behind it is basically pretty simple.

For example, several years ago I was injured in a boating accident and had to spend a week or so in the hospital. Over the next several months, I became anxious whenever I rode in a speedboat. That was because my brain had established an association between boats and the pain I had endured from the accident. Just sitting in a gently moving boat was enough to set off warning signals of the potential danger of another accident. Happily, my fears eventually disappeared—an improvement welcomed not only by me but also by my boating companions, who no longer had to put up with my periodic expressions of alarm in response to perfectly normal boating conditions.

In a more formal conditioning procedure, the experimenter places an animal, usually a rat, in a small cage. A sound then comes on, immediately followed by a brief electric shock to the animal's feet. It takes only a few pairings of the sound and the shock before the rat begins to act afraid whenever it hears the sound alone. At such moments, the rat stops whatever it's doing and freezes into a state of near perfect petrifaction—crouched down and motionless except for the tiny chest movements that accompany breathing. Freezing into such statuesque immobility is a universal response to threat in many species because it reduces the likelihood of being attacked.

But the conditioned response and the resulting freezing behavior can be extinguished if the sound continues to occur *minus* the electric shock. It's sort of like what happened to my fear of boats: After several uneventful boat rides, the association of a boat with an accident eventually disappeared.

Experiments on conditioned fear suggested to brain scientists a strategy for mapping out the fear-processing circuits within the brain. "All we have to do is trace the pathway forward from the input (the sensory system that processes the conditioned stimulus, the sound) to the output (the system that controls freezing or other hardwired responses)," wrote neuroscientist Joseph LeDoux in his book *Synaptic Self.* "The fear-processing circuits, by this logic, should be located at the intersection of the input and output systems."

Eventually, LeDoux identified the brain region that orchestrates the fear response. A small almond-shaped structure on both sides of the brain, the amygdala (or amygdalae when talking about both of them) contains more than a dozen distinct divisions. Two of the divisions are especially important.

Stimulation of the *central nucleus* produces many of the expressions of conditioned fear: heart rate increases, "freezing," and so on. Its destruction, in contrast, induces just the opposite reaction: disappearance of the conditioned fear response.

So much for the output system. What about the input system? How does a stimulus (foot shock) get to the amygdala in the first place?

A second nucleus in the amygdala, the *lateral nucleus*, functions as a receiving station for information flowing from the cortex and the thalamus, the two principal conduits within the brain for information coming from the outside world. "Thus the amgydala is able to monitor the outside world for danger," wrote LeDoux. "If the lateral nucleus detects danger, it activates the central nucleus which initiates the expression of behavioral responses and changes in body physiology that characterize states of fear."

Think of the amygdala as the hub of a wheel with spokes arriving along pathways from different parts of the brain. Since the amygdala appraises emotional meaning, an emotionally arousing situation, especially fear, activates the amygdala to stimulate the hypothalamus, leading to the release of hormones and other chemicals. The purpose of this response is to detect danger and mobilize a response. And since the amygdala also activates the hippocampus and other cortical areas involved in memory, the result is a stronger, more enduring memory of the dangerous situation.

Additional experiments in LeDoux's laboratory revealed the existence of a "low road" for the conditioned fear response. Figure 2 shows the circuits. The direct pathway from the thalamus to the amygdala is shorter and faster. As a result of

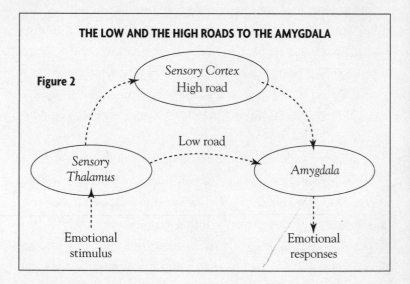

**THE LOW AND THE HIGH ROADS TO THE AMYGDALA**

Figure 2

Sensory Cortex
High road

Low road

Sensory
Thalamus

Amygdala

Emotional
stimulus

Emotional
responses

the activation of this pathway, the animal freezes at a loud sound prior to learning the source of the sound. The second pathway—from the thalamus to the cortex and then to the amygdala—allows additional time for the cortex to come up with an explanation for the sound.

Lest all of this strike you as unrelated to anything in your own life, consider that the proper functioning of the two circuits involving the amygdala can literally mean the difference between life and death. Since it takes about twelve milliseconds for a sound to reach the amygdala via the thalamus, and nearly twice as long to take the longer path through the cortex, you would be in grave danger if you were reliant strictly on the "high road" that carries information from the thalamus to the cortex and only then to the amygdala.

For instance, if you step off a curb and vaguely glimpse out of the corner of your eye something bearing down on you, your instant response is to step back onto the curb—thus avoiding being struck by that car that just ran the red light. Your narrow escape depended upon rapid transmission of the visual signal along the low road from the eye to the thalamus to the amygdala. If you had depended on the high road, chances are that the car would have hit you during the extra milliseconds required for the signal to go from the thalamus to the cortex, where you consciously appraised what was happening, and only then to the amygdala.

The low road provides what LeDoux refers to as a "quick and dirty" response that is often sufficient to save your life. Seconds later, when the danger is over, the signal coming to the amygdala from the cortex via that high road will provide the explanation for what happened. And it's at that instant that you'll first consciously experience fear, as you realize that if you hadn't acted as quickly and automatically as you did, you might now be dead or seriously injured.

Think back to that startle response I mentioned at the beginning of this chapter. The low road was the path responsible for my initial reaction to my sister's prank. My body responded as if there really *was* someone or something threatening me from behind. Seconds later, the information delivered along the high road informed me that it was only a prank. But since I was already aroused and frightened, I responded with the fear-anger-aggression response that typically follows startle. (Fortunately, my frontal cortex remained in command, and I was able to limit my aggressive response directed at my sister to a few angry words.)

Of course, we aren't aware of the brief interval that separates the arrival of information along the high and low pathways. As a result, we often attribute our fear or anxiety to the moment of our conscious recognition of the threat. But despite our impressions to the contrary, the fear or anxiety actually started earlier. We experience fright, and then we learn what it is that frightened us. And when the full impact of our experience really hits us, our heart starts to race, our breathing becomes labored, and we break out into a sweat. But our fear response actually began milliseconds prior to any conscious recognition of what was happening

AT THIS POINT, we're ready to explore the brain mechanisms underlying two topics we've previously described: the psychological and behavioral differences between fear and anxiety; and the startle response. First, let's examine startle in more detail.

If you hear a loud and unexpected sound, your eyelids will blink involuntarily (the same reflex also occurs in response to an unexpected puff of air to the eyes). This acoustic startle reflex is the same one that triggered the rapid flexion and extension of muscles throughout my body that led to my "collapse" in response to the fright caused by my sister's prank. In most cases, the response is subtler than what I experienced— largely limited to those rapid, almost imperceptible eye blinks. In either case, the reflex can be modified in interesting ways.

Imagine that we're observing a rat in a cage. Over the previous several weeks, this rat has served as the involuntary participant in the following experiment: After turning out the lights and placing the rat in the cage, the experimenter briefly turns on a light, followed three to four seconds later by a shock. After a small number of pairings of the light and the shock (the number varying from one rat to another), the animal's startle response to a loud sound increases if it occurs three to four seconds after the light. No mystery here. The rat is frightened when the light comes on because it expects it will get a shock. As a result, the startle reflex increases at the loud sound just as it did at the shock. This so-called fear-potentiated startle effect—an example of Pavlovian conditioning—can also be elicited from humans.

Now change the experiment so that light comes on repetitively in the absence of the shock. When the rat's startle reflex to a loud sound is tested again, the light now fails to increase the startle response—a sign of the extinction of the previously established fear conditioning. So much for the fear response. How would that differ from an anxiety response? And where do fear and anxiety originate in the rat brain?

To answer those questions, I have to give you some more information about rats and light. The creatures are sufficiently frightened of illumination that their tendency to startle increases and remains increased even when they're exposed to illumination outside of a conditioning situation (no loud noises or shocks; just a bright light for anywhere from five to twenty minutes).

Just the opposite would occur if you or I participated in a similar experiment: Our general uneasiness and tendency to startle would be greatest in the darkness. And when you think about it, this species difference makes sense: Light is threatening to a nocturnal creature (the rat), while a diurnal one like ourselves becomes edgy in the darkness. In response to our anxiety, we calm our fears by turning on lights, thus freeing ourselves from the hobgoblins created by our imagination. That's also why a scary movie is always scarier when we watch it in the darkness of a movie theater than when we watch it at home in the brightness of a sunroom. (As far as we know, rats don't watch scary movies; but if they did, they'd prefer to view them in daylight in order to get the maximum scary effect.)

Equipped with this information about a rat's responses to light, let's explore the distinction within the brain between fear and anxiety. Once again, we return to the centerpiece of the fear system, the amygdala.

If the central nucleus of the amygdala is destroyed, a rat can no longer be conditioned to react with an increased startle response to the pairing of a light with a shock. It will continue, however, to show a light-enhanced edginess and a tendency to startle in response to a light turned on and kept on for anywhere from five to twenty minutes. It's as if the conditioned fear mediated by the central nucleus has disappeared following amygdalar destruction, while the animal's general background anxiety remains: While the light remains on, the rat remains poised in anticipation of something threatening about to happen. So where in the rat brain is this anxiety response located?

After more than twenty years of meticulous anatomical research, Michael Davis, a neuroscientist at Emory University, has discovered that fear and anxiety in the rat are mediated by different brain areas. While fear results from central nucleus stimulation, anxiety is processed in an area closely linked to the amygdala called the bed nucleus of the stria terminalis. Destruction of this area does away with the rat's free-floating discomfort in daylight (a sustained reaction just like anxiety) but leaves intact the rat's development of conditioned fears, such as a flash of light paired with a shock.

"The bed nucleus of the stria terminalis responds to signals more akin to anxiety than those akin to fear, whereas the central nucleus of the amygdala is involved in fear and not as much in anxiety," wrote Davis.

Although no one so far has proven that Davis's research on rats is applicable to humans, it is probable that fear and anxiety are also encoded differently within our brains. As a practical consequence, tranquilizers may be appropriate treatment for anxiety—but realistic fears shouldn't be medicated away.

For example, I recently encountered a journalist who came to my office and told me of an upcoming overseas assignment that would require him to fly in a small plane over desolate and hostile terrain. He explained that the pilot assigned to the project had a reputation for risky, even careless, maneuvering. Based on several near accidents when flying with that pilot in the past, the journalist had become convinced of the pilot's poor judgement. He requested that I write him a prescription for a tranquilizer to quell his "nervousness" when flying with this pilot. His dilemma reminded me of the need to preserve this important distinction between fear and anxiety.

I told him that his response was appropriate under the circumstances and that he should insist that his employers hire another pilot. What was needed instead of a medication, I told him, was his recognition that his fear was well founded and required a respectful but firm insistence on a pilot change. I reminded him of Edmund Burke's aphorism "No passion so effectively robs the mind of acting and reasoning as fear." In the journalist's case, the fear was so disturbing that he felt the need to suppress it rather than recognize that under the circumstances, fear was an appropriate, even lifesaving response to a truly dangerous situation. He took my advice and insisted on a new pilot. I later learned that his fear disappeared. In his case, since fear served as a motivating force warning of imminent danger, an antianxiety medication would have been ill advised and even dangerous.

The amygdala also plays an important role in determining our responses to less-threatening situations. For example, if I show you a picture containing various objects (a car, a key, a book) and ask you to quickly locate, say, the key, the speed of your response will, not surprisingly, vary with the number of objects in the picture. The greater the number of objects, the more time you will require in order to find the key—a reflection of the fact that you had to visually scan everything, one object a time, until you locate the key. But if, instead, I show you a picture in which a more emotionally arousing object is substituted for the key, you will respond quite differently. Snakes or spiders will be homed in on more rapidly than cats, dogs, or other less-threatening animals. The more fearful the image, the quicker and more efficiently your brain will process that image. This "capture of attention" effect, as researchers refer to it, is especially strong in regard to faces. Grimacing faces, for instance, will be detected more quickly than faces with neutral expressions.

Recordings made during tests like I've just described reveal something even more intriguing: Thanks to the amygdala, we respond to emotional facial expressions even before identifying the stimulus as a human face! Ordinarily, identification of a face coincides with a face-related electrical response (measured by a technique called magnetoencephalography, or MEG) occurring 170 milliseconds (thousandths of a second) after the appearance of the face. Responses to faces showing emotional expressions, however, occur as early as 100 milliseconds.

Neuroscientists uncovered other intriguing findings when testing brain-damaged patients deprived of vision off to one side. After damage to the right inferior parietal cortex, an area on the right side of the brain located several inches above the ear, the affected person frequently loses the ability to perceive or respond in any way to things in the opposite (left) visual field. But this doesn't hold for angry faces or images of spiders or other anxiety-arousing material. Despite the affected person's usual inability to consciously see and describe anything on the blind side, his or her amygdala responds when the experimenter projects an emotional picture to that side—evidence of a split between conscious and unconscious perception. And when I speak of unconscious perception, I'm not referring to the sex- and violence-ridden unconscious espoused by Freud, but the unconscious of contemporary cognitive psychologists, in which brain processing occurs outside of awareness.

One more important point about the amygdala: It plays an important role in memory and learning. Think back to where you were and what you were doing when you first learned about the terrorist attacks on New York and Washington on September 11, 2001. The clarity and precision of your memories for that September 11 date result from the activation that took place in your amygdala at the moment you first learned of the attacks. An fMRI of your brain taken at the time would have shown enhanced activity in one or both amygdalae as a result of the highly upsetting and arousing nature of the events. You weren't responding to the attacks that day as you would to an emotionally neutral experience. Instead, you were highly aroused. That enhanced emotional arousal resulted from increased firing from your amygdala, which endowed that memory with such vividness that today you can clearly remember the details surrounding your learning about the attacks.

Your memory of September 11, 2001, is based on the extensive two-way connections linking the amygdala to the visual cortex and the hippocampus. Thanks to these linkages, most of us can remember more about September 11, 2001, than about events occurring on that same month and date of any other year. Indeed, the emotional impact of that event will probably never completely recede; each time you think about it, you will activate your amygdala once again.

## A SUGGESTION FOR TRANSFORMING YOUR ANXIETY

*Learn as Much as You Can About What's Happening
in Your Brain When You're Feeling Anxious*

Learning about the brain mechanisms involved in anxiety provides a new way of experiencing and thinking about it. The more you learn about the neurobiology of anxiety, the more objectivity you can bring to the task of managing your own personal anxiety. And as you learn more, you will be in a better position to transform your anxiety from something to be feared to something you can critically examine and even use to your advantage. So, the next time you're feeling anxious, think about the brain circuitry that underlies your anxious responses: the role of the amygdala, the conditioning responses, and, most of all, the power of the frontal lobes to override or at least moderate the influence of the amygdala and other components of the emotional circuitry. Remind yourself that anxiety isn't just something that "happens"; it has an underlying physical basis within your brain. Therefore, learn as much as you can about the brain mechanisms responsible for anxiety. Reread the detailed descriptions of the brain's anxiety circuits I've provided in this book; consult the other sources I've referenced in the bibliography.

# 5

# Nobody Messes with Dave
*The Experimental Manipulation of Anxiety*

URING A BRAIN RESEARCH meeting in San Francisco, I caught up with David Amaral, who edits a scientific journal named *Amygdala* and works at the Center for Neuroscience at the University of California at Davis. Amaral has been carrying out research on the amygdala over the past two decades, currently as director of the California Regional Primate Research Center. His research is focused on the effects of amygdalar lesions on rhesus monkeys selected from a colony numbering more than four thousand.

Before describing Amaral's research, I think it's important to repeat that this is not a book about the rights or wrongs of animal experimentation. It is about anxiety. Unfortunately, much of what we've learned about anxiety has resulted from surgical experiments carried out on animals. If reading about animal experiments disturbs you, you might want to skip the next six pages, although I think you miss a lot if you do.

Amaral's interest in the amygdala began in 1954 after he read a research paper describing the effects of bilateral removal of the amygdala on the dominance hierarchy among rhesus monkeys. Ordinarily, monkeys live in communities in which some monkeys give the orders and some take them—communities, in other words, much like our own. Within this hierarchy, one member of the monkey colony is the boss, the chief honcho that controls and dominates the other monkeys. But that honcho can be transformed into a wimp as the result of damage to the amygdala.

A cartoon taken from the 1954 paper depicts a colony of eight monkeys aligned in a dominance hierarchy, starting at the top of the chain with Dave, described as "dominant, feared and self-assured." And judging from the cowed and submissive body language adopted by the other monkeys lower down on the dominance chain, it's obvious that nobody messed with Dave. But things changed drastically after the surgical destruction of the amygdalae on both sides of Dave's brain. A second cartoon drawn after the operation depicts a different Dave, one that now dwells not at the top but at the bottom of the hierarchy. The cartoon shows Dave running from a pursuer, "an outcast fleeing all," as described in the caption.

Based on this early experiment on Dave, Amaral decided to dig a bit deeper into the mysteries of the amygdala. Why would destruction of this tiny brain region lead to such a dramatic alteration in a monkey's hierarchical position? To find out, he modernized the 1954 operation by carrying out more precise brain alterations. He discovered that after the operations, the monkeys showed less fear when confronted with a snake or other ordinarily fear-arousing stimuli. They also shed the usual wariness monkeys characteristically adopt when first encountering a strange monkey.

"When normal macaque monkeys that haven't had previous contact are introduced into a common space, they take a considerable amount of time to evaluate each other," said Amaral. "Typically, the monkeys remain aloof and avoid getting too close to each other—a behavioral pattern that decreases the likelihood of aggression. But after getting to know one another a bit, the monkeys begin interacting. In contrast to this pattern, amygdala-damaged animals begin engaging in social activities at the moment they are placed with the new animal."

In short, the animal lacking both amygdalae loses the natural and potentially lifesaving tendency to stand back a moment and survey each unfamiliar monkey as a potential threat. In another test of these behavioral differences, Amaral used what's known as the Human Intruder paradigm. This test was based on the tendency for normal monkeys to attack or otherwise react with hostility whenever a human looks them directly in the eye. Monkeys lacking their amygdalae, in contrast, don't get upset—they simply stare back.

Nor do they respond to things that frighten normal monkeys. Amaral showed me a video of a normal monkey running away when confronted with a plastic snake followed by a second video of another monkey after an amygdalectomy (operation on the amygdala). That second monkey showed no fear of the snake; rather, he picked it up and stared at it for several moments.

"The amygdala acts as a protection device," according to Amaral. "It is designed to detect and avoid danger. If the animal encounters a predator such as a snake, the amygdala assesses the snake as a threat. Indeed, the danger of the stimulus is only appreciated once the amygdala is involved. It functions as a protective 'brake' on engagement with other animals or objects until a threat assessment is carried out."

Based on these initial research findings, Amaral wondered: Suppose the amygdala operation were done on infant instead of adult macaques? Would the consequences be any different? To find out, he carried out a small number of experiments involving an operation on the amygdalae of two-week-old animals. He then returned the animals to their mothers.

Over the next several weeks, the mutual responses of infant and mother appeared to be perfectly normal. Also similar to their adult counterparts was the response to inanimate objects like rubber snakes: The infant monkeys whose amygdalae had been destroyed picked up the faux snakes with no hesitation whatsoever. So far, no surprise: Lesions in infant and adult macaques produced the same results.

The surprise came after weaning, when the animals first encountered potential playmates. Instead of prematurely schmoozing and taking on the role of the life of the party (the response of the adult monkeys after operations performed on their amygdalae), the monkeys that had undergone the amygdalae operation as infants turned fearful, stayed to themselves, and communicated in no uncertain terms that they wanted to be left alone.

I asked Amaral about this apparent contradiction between the adult's and the infant's responses when encountering unfamiliar monkeys: Destruction of the amygdalae in the newborn monkey, in contrast to the same operation in the adult monkey, resulted in *increased* rather than decreased fear in social situations.

"It's less contradictory if you think of the amygdala as essential for gaining social knowledge, especially interpreting and acting on social signals from others," he said. "From this point of view, damage to the amygdala will have a greater effect early in life because the structure hasn't had the opportunity to incorporate social knowledge. Thus the fear shown by the infants to strange monkeys may result because they approach complex social situations without the necessary knowledge about how to act appropriately. Those monkeys that sustained amygdala damage as adults, in contrast, have already learned and can apply some of the rules for behaving around other monkeys."

Amaral's work suggests the existence of a dual fear system: One system handles fear of inanimate objects, the other handles social fear. Such an arrangement should not be surprising. Most creatures, including ourselves, usually find social interactions more complicated and fraught with anxiety than we do interactions with inanimate objects. Context is also important. Staring at a snake in a cage across from you in the San Diego Zoo will elicit a different response than encountering that snake in the desert in Arizona. That's because the snake in the zoo isn't cause for alarm; your reasoning powers remain firmly in control ("I'm protected by the glass, so I can relax"). But when you encounter that snake in the desert, the message is urgent ("Run before the snake attacks!") and marked by an enhanced amygdalar response.

While Amaral cautions against directly comparing his research to that on humans, he cautiously speculates that two anxiety disorders, social anxiety and social phobia, might be related to hyperactivity of normal amygdala function: "If the set point for determining the dangerousness of an object, an individual, or a situation was set below what is normally ben-eficial to the individual, benign people and objects might be judged dangerous and avoided."

Amaral's animal research is certainly consistent with such a hypothesis. Functional MRI studies in humans show left amyg-dala activation in the presence of threatening situations. The research also jibes with the finding that it is difficult to inter-pret facial expressions of fear in patients lacking a normally functioning amygdala. Overall, the cost of an abnormally functioning amygdala both in the monkey and the human can range from the merely inconvenient to the fatal: In the pres-ence of the snake, the proper and life-sustaining response is not curiosity but fear.

But a fearful facial expression requires a more subtle response. When I correctly "read" another person's fearful facial expression, my perception fails to answer an important but more ambiguous question: What is it that the person is fearful about? In other words, my encounter with a fearful face activates my amygdala because of the basic ambiguity of what I'm seeing: The other person's fearful facial expres-sion may result from any number of things in the immediate environment.

Perhaps it's best, therefore, to think of the amygdala as an *evaluator* of situations that may potentially pose a danger. Such evaluations, of course, will vary from one person to another. While one person may cringe in horror at the sight of a spider in the kitchen, another, experiencing only annoyance, will impatiently reach for the nearest magazine to crush the unwelcome visitor. Based on the research of Amaral and others, it seems likely that such differences in the evaluation of dangerousness result from variations in normal amygdala function.

BEFORE WE FINISH with the amygdala, let me tell you about a dramatic experiment that illustrates how the powerful influence of the amygdala can be switched on and off within the brain of a single test subject. J. L. Downer of University College, London, removed the amygdala on only one side of a monkey's brain (along with the creation of other brain modifications incidental to my point). As a result of this surgical procedure, a kind of magic trick can be performed featuring the remaining amygdala.

You start the trick by covering the monkey's eye on the side of the intact amygdala. As a result, visual input from the eye on that side no longer flows to that amygdala. Absent that connection between the eye and the amygdala, the monkey turns placid, even mellowing out enough to step forward and take a raisin from your hand. But if you then switch the blindfold to the other eye so that the monkey now looks out upon the world through the eye connected to the intact amygdala, jungle rules apply: The same animal will scratch, bite, and otherwise go on the offensive. Finally, if you once again cover the eye reporting to the intact amygdala, the nastiness subsides and the creature once again wants to make nice. In short, the amygdala is the creator and activator of the emotional oomph that makes a monkey a monkey. And if you deprive the animal of the contribution from the amygdala, the monkey turns as tame as a pampered pussycat.

Obviously, it would be totally unacceptable except in a sci-fi movie to try a similar experiment aimed at deliberately damaging the amygdala of a human. But sometimes nature carries out experiments on its own that no scientist would ever attempt. Several years ago neurologist Antonio Damasio and his associate Daniel Tranel, both from the University of Iowa College of Medicine, encountered a young woman identified in their professional papers simply by the initials SM. She was unfortunate enough to have suffered severe damage restricted to the amygdala on both sides of her brain.

In order to test SM's responses, Damasio and Tranel designed a painless conditioning experiment. They discovered that she could not be conditioned to the combination of a light and (of all things) the recorded blast of a boat horn. In the conditioning experiment, several different colored lights were shown to her in varying sequences, but only the appearance on a screen of a blue light—not a light of any other color—preceded the horn blast.

In people with normal brains, such a linking of a color with an unpleasantly loud noise induces a fall in the electrical resistance on the skin of the palm (the result of sweat gland stimulation that increases the amount of sweat, an excellent conductor of electricity). In technical terms, conditioning can easily be established in the normal person by linking the conditioned stimulus (the light) and the unconditioned stimulus (the blast). But conditioning could not be established with SM: No alteration occurred in the electrical resistance measured from her palm. Nonetheless, she had no problem concluding on the basis of her observations that only the blue light portended the imminent blast of the horn.

Nor were SM's problem's restricted to the circumstances of a somewhat restricted scientific experiment. (When was the last time you heard a blast from a boat horn just after looking at a blue light?) In a more lifelike experiment, SM looked at pictures of actors posing with the facial expressions characteristic of different emotions (fear, anger, disgust). She experienced great difficulty recognizing and identifying the appropriate emotions corresponding to the various facial expressions. She was especially impaired in recognizing facial expressions of fear. With the exception of this inability, SM's performance was otherwise normal: She had no problem distinguishing one person from another, even with pictures taken from different angles.

A second experiment on another person unfortunate enough to have lost the use of both amygdalae turned up additional emotional oddities. This person not only had problems identifying facial expressions of emotion, but he also couldn't recognize the vocal intonations customarily used to convey and experience emotion ("Could you *please at least try* to be on time just this once?"—that kind of thing). Not only could this patient not recognize tonal patterns in voices that convey emotion, but he also experienced difficulty distinguishing statements, questions, and exclamations on the basis of changes in the speaker's inflections. Yet a third patient with bilateral amygdala damage experienced problems in tests of social judgment, such as deciding whether a person appears trustworthy.

In normal volunteers, in contrast, emotional expressions immediately activate the amygdala, even when the subjects fail to consciously register the emotion-arousing stimulus. For example, if the experimenter briefly flashes an emotional face on a screen and then quickly follows up with a face that appears emotionless, the amygdala fires up even though the subject consciously perceived only the face with the neutral expression.

Fortunately, the brain exerts a check on the influence of the amygdala. And for good reason. Just imagine the problems that would arise if our amygdala encountered no limits on its ability to capture the brain's attention. Each time we encountered something frightening or anxiety arousing, we would automatically respond no matter what the circumstances: Fight or flight would become the operative principle governing our lives. Indeed, that is what happens to people afflicted with anxiety disorders.

Under normal circumstances, the brain oversees the total situation and responds in a measured way. When we're watching a horror movie, for instance, our amygdalae activates, yet we don't cut and run for the exit. Rather, we enjoy the vicarious sensations of fear or anxiety. We can do this because a part of our brain, the frontal lobes, "whispers" to the rest of the brain, "This isn't real, but only make-believe." And since the frontal lobes play such a pivotal role in our mental lives, especially in the management of anxiety, let's take a closer look at them.

# 6

# Balancing Act
*The Amygdala and the Frontal Lobes*

I HAVE A FAVORITE slide I like to show during my lectures on the brain. It depicts the size differences in the cerebral hemispheres among creatures ranging from rats and moles and extending upward to macaques, chimpanzees, and, finally, *Homo sapiens*. Not only are the cerebral hemispheres of the human brain larger than in any of these other species, but the frontal and prefrontal lobes, the areas located just behind the forehead, can't be clearly demarcated in creatures other than ourselves. Over the past hundred years or so, neuroscientists have honed in on the contributions to our mental functioning provided by those enlarged frontal lobes.

Our frontal lobes enable us to construct inner representations of the outside world. Thanks to our frontal lobes, we can imagine what might happen if we carry out certain actions or fail to carry out others. Thus, we can easily bring to mind the likely financial and legal consequences of, say, not filing our income tax returns.

In short, our frontal lobes help us to foresee threats before they materialize, and that's a good thing. For instance, when we're walking down a dark alley, our frontal lobes prod us into scanning our surroundings with the intention of turning back or picking up the pace if we see or hear anything potentially menacing. But sometimes the frontal lobes also contribute to making us fearful when there isn't anything to be afraid of. And that's not such a good thing. In those situations, our imaginations do us in.

Thus, the frontal lobes can prove either an asset or a liability, depending on the circumstances. On the asset side, this most advanced portion of our brain enables us to mentally step back from our immediate circumstances, foresee the potential consequences of our actions, and plan accordingly.

But overactivity of the frontal lobes can prove a liability by spinning elaborate fantasies about improbable, even impossible, occurrences. The result can be agonizing obsessions about all of the terrible things that could conceivably happen if certain thoughts aren't kept in mind, specific phrases spoken, or time-consuming rituals compulsively carried out. Obsessive-compulsive disorder (OCD), which we will have more to say about later, results from a disturbance in the circuit linking the frontal lobes with lower brain centers within the basal ganglia. Overactivity in this frontal-basal ganglia circuit can be seen in PET scans of people afflicted with OCD. You can therefore think of the frontal lobes as the elaborator of the thoughts and ideas that provide anxiety themes. We know this on the basis of a grisly operation (read unethical human experiment) known as prefrontal lobotomy, which provided macabre proof of the importance of the frontal lobe in generating anxiety.

In the 1940s—a decade prior to the availability of tranquilizers—some psychiatric patients drove themselves and everybody around them to desperation thanks to their extreme and uncontrollable anxiety. They paced about for hours at a time, beseeched help from family and friends, and wrung their hands in desperation at the mental tortures they were experiencing. Not much could be done at the time to ease their torture. They could be hospitalized and placed in seclusion, rendered temporarily unconscious by means of intravenous barbiturates, or simply left to suffer. But all of this changed with the advent of prefrontal lobotomy—sometimes performed in a doctor's office rather than in a hospital operating room—which severed the nerve fibers projecting from the frontal lobes. After prefrontal lobotomy, the patients no longer experienced anxiety.

Although the operation successfully relieved anxiety, it created different but no less disturbing problems. Following lobotomy, the patients lost all interest in the people and events going on around them; they spent long hours simply staring into space. Relatives and friends commented that the operation had obviously gone too far; it had eliminated not only anxiety but also the basic components of the patient's personality. Docile, intellectually retarded, and emotionally unavailable, the patients lacked the capacity to focus their attention, resist distractions, or emotionally connect with other people. Indeed, they bore an unsettling resemblance to the zombies depicted in science-fiction movies popular at the time.

As a result of the discomfort most people experienced when dealing with lobotomized patients, they tended to shun them— usually by placing them in mental hospitals, where many of them remained for the rest of their lives. Fortunately, the abuses of psychosurgery eventually came to public attention and the operation fell out of favor—but not before the performance of this savage and unregulated procedure on somewhere between ten thousand and fifty thousand Americans.

The lobotomy experience provides dramatic proof of the human costs that result from attempts to totally eliminate anxiety. A more reasonable and helpful goal is to understand the relevant brain mechanisms responsible for it. And recent research is making a significant contribution toward this goal by demonstrating the intimate interrelation between thoughts and emotions. In the next few paragraphs I want to tell you about some of that research.

Imagine yourself staring at a computer screen while waiting for the appearance of a green, red, or blue light. Immediately following the green light, you feel a mildly pleasant heat sensation delivered to the back of your hand via an attached metal plate. A red light then briefly comes on, followed by a blue light. Just after the appearance of the blue light, you feel an unpleasantly hot sensation on the back of your hand. After only a few trials, your brain will learn the association between the colors of the lights and the prospect of imminent pain.

In this fMRI conditioning experiment carried out by neuroscientists at Oxford University, the brain's responses during the anxious period prior to the pain differs from the responses occurring during the pain. (The activation during the period of anxiety involved, not surprisingly, limbic system structures.)

In fact, this distinction is sufficiently precise that anyone look-ing at the fMRI images would be able, after only a few trials, to identify those moments when the subject is anxiously antic-ipating the pain. In short, fMRI now makes possible a clear distinction between an event and anxiety about that event.

Our level of anxiety can also be correlated with our ability to stay focused. You can probably recall occasions when you experienced difficulty preparing for an examination because you were overly anxious that you wouldn't do well. As a result of your anxiety, you found it hard to concentrate and focus on your notes, or even to sit still long enough to read them.

In a study of the effects of anxiety on a person's ability to stay focused, researchers from Duke University requested vol-unteers to keep track of a series of visual targets displayed on a screen in front of them. While the subjects attended to the circles on the screen, the researchers distracted them with images that were either neutral (people sitting around a table) or likely to evoke fear or anxiety (a car accident or a scene taken from a trauma center).

Only the anxiety-provoking images proved distracting. Moreover, the functional MRI images taken of the subjects during the experiment showed emotions and attention moving in parallel streams through the brain before finally coming together in the anterior cingulate, a structure located between the right and left halves of the frontal lobes. This tiny struc-ture helps to create a balance between emotions and concen-tration: An increase in one type of function is accompanied by a noticeable decrease in the other.

Such findings are very much in line with our subjective experiences: Emotions interfere with our ability to focus and attend. For instance, when we're trying to concentrate, anxiety leads to distraction, as in the examination situation mentioned above. Whatever the challenge, inattention and distraction worsen our performance.

Now, since emotion and concentration flow along two separate pathways, could an anxious person counter the effects of his or her anxiety by enhancing that second pathway to the anterior cingulate through more determined efforts at paying attention? Such a possibility is certainly in line with the suggestion I give to my anxious patients that instead of allowing their anxiety to remain at the center stage of their consciousness, they should occupy their minds, immerse themselves in some activity other than passively give in to their anxious feelings.

Despite the slightly Pollyannaish flavor of such advice, most of us can recall occasions when we were caught up in anxious anticipation of some upcoming event, only to find some measure of relief by "distracting" ourselves with an engrossing activity (a misnomer, since we were actually *focusing* ourselves in face of the distraction caused by our anxiety). I mention this approach because I don't want to leave you with the impression that the research I've mentioned in this chapter implies determinism—that the discovery of networks and pathways within the brain responsible for the mediation of anxiety places insuperable limitations on our freedom. Instead, we retain the freedom at any time to employ our frontal lobes to switch our mental focus from anxiety to concentrated attention.

One final point about anxiety and the anterior cingulate. Neuroscientists have recently discovered abnormal firing patterns in response to both physical and psychological pain. In my favorite experiment along these lines, experimental psychologist

Naomi Eisenberger of UCLA placed several subjects in an fMRI and scanned them while they played a virtual ball-tossing video game (Cyberball) with what they believed to be two other players in nearby fMRI scanners. Actually, there were no other players; a preset computer program provided all the responses.

In the first phases of the experiment, Eisenberger told her subjects that "technical difficulties" precluded them from participating in the game, but that they could watch the action between the other "players." Although disappointed, the subjects waited patiently. A few minutes later, after the faux technical difficulties had been "fixed," the subjects joined the game. But soon after the subjects began participating in the video game, events took on a distinctly unfriendly turn: the other "players" soon stopped directing throws to the subjects. Not surprisingly, the subjects reported feelings of distress. Accompanying their distress at what appeared to be a purposeful exclusion of them from the game was fMRI evidence of increased activity in the anterior cingulate. No such pattern had appeared during the subject's earlier exclusion because of "technical difficulties." In short, Eisenberger found that it made a great deal of difference whether the exclusion resulted from unforeseen circumstances or the deliberate action of the other "players."

"In essence, a pattern of anterior cingulate activations similar to those found with physical pain emerged during social exclusion, suggesting that the experience of social and physical pain share a common neuroanatomy," said Eisenberger. But would it be fair to equate this with anxiety?

"Of course," she responded to my question. "In fact, I first became interested in the cingulate cortex because I was interested in anxiety. Individuals with anxiety tend to have more activity in the anterior cingulate, so it's fair to say that the anterior cingulate is involved in separation distress/separation anxiety."

In summary, the anterior cingulate serves as a barometer of anxiety, especially anxiety resulting from perceived social rejection. So the next time you're upset because someone ignores you, remind yourself that your feelings are the result of your anterior cingulate going into overdrive. Further, the amount and intensity of the anxiety you're experiencing will vary according to the balance that exists between your frontal lobes and your amygdala. Too much activation of the amygdala, and you'll respond overemotionally; just the right amount of activation from the frontal lobe, and you will be able to put your anxieties into perspective.

But suppose your frontal lobes aren't functioning normally. What effect would that have on your anxiety? An answer to that question comes from electrical stimulation experiments on a part of the frontal lobe called the orbital cortex. Stimulation of that area in animals (and in humans as well if the area is electrically prodded during a brain operation) increases blood pressure, heart rate, and breathing. These physiological processes are, of course, the very bodily responses that accompany fear and anxiety.

Think about what usually happens when you're about to make an important decision. Your pulse and respiration rates increase and you feel anxious. These changes wouldn't occur if part of your prefrontal cortex were damaged. Nor would you experience anxiety. Indeed, if you were deprived of the influence of the prefrontal cortex, you would no longer experience anxiety in situations when anxiety is appropriate. We know this not only on the basis of the lobotomy operations mentioned earlier, but also based on an ingenious experiment carried out by Daniel Tranel, the director of the neuroscience program at the University of Iowa College of Medicine.

IN THE 1990s, DANIEL TRANEL and his colleagues designed a game they named the Gambling Task. Here is how it is played: After receiving a start-up loan of $2,000 in play money, two players select one card at a time from any one of four decks. After each card selection, the subjects receive money, the amount varying from deck to deck. But after some card selections, the subjects are both given money and asked to pay a penalty; again, the amount varies from deck to deck. Throughout the game, the players are free to switch from any deck to another, at any time and as often as they wish. After the players have selected one hundred cards from any combination of the four decks, the game ends. As with any gambling game, the goal is to win as much money as possible.

Unknown to the subjects, two of the decks are "fixed" so as to yield higher immediate rewards, but also higher long-term penalties. Thus, the player who selects cards only from these decks will initially accumulate more money than if he or she had selected from the other two decks. But as the game proceeds, the penalties associated with selecting from the fixed decks increase to the point that the player suffers a net long-term loss. The other two decks, in contrast, yield lower immediate rewards (less money is initially paid out), but more money is earned in the long run, since selecting cards only from these decks leads to fewer penalties and a net long-term gain by the conclusion of the game.

In order to come out ahead in the Gambling Task, a player must eventually recognize that he or she should select cards from only two of the four decks. Typically, this insight doesn't occur immediately; rather, it requires periods of trial and error that vary from one player to another. Initially, most players sample all four of the decks and, prior to encountering any of the punishment cards, tend to limit selection to the fixed decks (in response to the higher immediate rewards). Tranel refers to this as the *pre-punishment period*. But as the players draw increasing numbers of punishment cards from the fixed decks and their initial winnings start to dwindle, fascinating responses occur that vary depending on whether the player's prefrontal cortex is normal or damaged.

Players with a normal prefrontal cortex gradually shift their card selections toward the normal decks (the "advantageous" decks, as Tranel refers to them) based on their gradually accruing losses from the fixed decks. At first, they can't identify the problem, other than that there is "something wrong" with the fixed decks. In response to their intuition, as the game progresses they tend to select only from the advantageous decks. When questioned about this change in their selections, most of the players initially can't come up with an explanation. But halfway through the experiment, the players begin to express a "hunch" that some decks are riskier. Finally, toward the end, most of the players voice a conviction that some decks are "bad" and others are "good."

The players with prefrontal damage, in contrast, persist throughout the game to favor the fixed (disadvantageous) decks even though they are losing money. What's going on here?

In order to answer that question, it's necessary to tell you one more important thing about Tranel's experiment. Along with the decks, he used instruments capable of measuring changes in his subject's nervous system. Ordinarily, emotions evoke bodily changes that can be measured and, on occasion, even observed without any special training or equipment. Think back to an occasion when you were embarrassed. As part and parcel of your embarrassment, you may have felt "flushed" at around the same time that others observed that you were blushing. This mix of private and public reactions resulted from the dilatation of blood vessels in your face, which caused an increase of blood flow and the subsequent flushing and blushing. But other changes were taking place as well.

For one, your emotional response led to an increase in the amount of fluid secreted by your sweat glands. And although the increased sweating may not have been noticeable either to you or to anyone shaking hands with you, even a small increase in sweat is sufficient to alter the passage of a low-voltage electrical current between a pair of detectors placed on the skin. Any change in the amount of current conducted (the skin conductance response, or SCR) takes the form of a measurable wave with a specific amplitude and frequency.

Tranel measured SCRs during the Gambling Task and found a regular recurring sequence among normal volunteers. First, reward and punishment SCRs occurred immediately following either the subject's selection of a card earning a reward or a card earning a reward along with a heavy offsetting penalty. Second, some SCRs, termed anticipatory SCRs, occurred immediately prior to the subject's selection of a card from one of the decks—that is, at the moment when they were deliberating about which deck to choose from. As the game progressed, the anticipatory SCRs became especially pronounced whenever the subject pondered whether to select a card from one of the fixed decks. In contrast, reduced SCRs appeared prior to the selection of a card from the ordinary decks. This difference in the SCRs prior to card selection from the fixed versus the legitimate decks suggests an early appreciation on the physiological level—prior to conscious recognition—that choosing from the fixed decks increased the risks.

Among subjects with prefrontal brain damage, however, the reward and punishment SCRs followed a different pattern. Although the subjects generated perfectly normal SCRs in response to reward and punishment, they failed to generate anticipatory SCRs. "This pattern of generating SCRs prior to their card selection never developed in the participants with frontal lobe damage," said Tranel. "This absence of anticipatory SCRs serves as a marker for their insensitivity to future outcomes."

The Gambling Task illustrates the importance of nonconscious processes in determining our behavior. Each of the normal volunteers began generating brain responses (as measured by SCRs) long before achieving conscious insight into the existence of the trick decks. Further, these responses eventually guided them to restrict their selections to the normal decks and avoid the trick decks. Think of the early SCRs as corresponding to an anxious apprehension of something amiss. With additional generation of SCRs, the players edged closer to the pivotal insight that certain decks must be avoided.

Finally, that hunch was confirmed when the players became consciously aware of the "rules" of the game.

Players with frontal lobe damage, in contrast, failed to generate anticipatory SCRs at any time during play. As a result, they continued to make selections that ultimately bankrupted them. In essence, their failure to develop an appropriate anxiety put them at risk of suffering serious consequences (or certainly would have if they were playing with real money and if Dan Tranel was a dishonest gambler instead of a brilliant and innovative brain researcher).

Tranel's gambling experiment provides another illustration of one of the basic themes of this book: Anxiety isn't always bad. "Too little emotion may be just as bad as excessive emotion," according to Tranel. "The Gambling Task suggests that nonconscious biases guide our reasoning and decision making. And without the help of such biases, conscious knowledge on our part may not be sufficient to ensure that we act to our best advantage."

During my discussion with Tranel, I tried to imagine what it would be like to go through life without any anxiety at all, to respond coolly and calmly to whatever occurred. We tend to romanticize such people as heroes, cheer for them in action movies, and fantasize about them in our daydreams. But while the sangfroid of the utterly anxiety-free person seems appealing, the end result of such a personality would likely prove catastrophic.

Consider this description by Dan Tranel of a man—let's call him Jim—who had undergone brain surgery during infancy for the removal of a tumor in his right frontal lobe. After the surgery, Jim's behavior didn't differ in any notable way from that of other children the same age. The trouble started when he began school. He couldn't adjust to new situations and needed continual reminders in order to keep focused on his work. A constant disciplinary problem, he disrupted classes and frequently failed to complete assignments. After eventually graduating from high school, Jim's behavior deteriorated even further. After being fired from several jobs, he remained at home watching TV or listening to music. Inept at managing money, Jim ran up credit card debts he couldn't pay and had no intention of paying.

According to Dan Tranel, Jim was "impulsive and exercised poor judgment. He lied frequently and maintained no lasting friendships. His parents described him as showing little worry, guilt, empathy, remorse or fear." Nor did Jim show signs of anxiety or ever complain of anything suggesting an anxious response.

In an attempt to diagnose the cause of Jim's condition, the neurologists working with Tranel did an MRI of Jim's brain. It revealed extensive damage in the right frontal lobe. And when participating in the Gambling Task, Jim persistently selected from the loaded decks. What's more, he failed to generate anticipatory SCRs prior to his selections.

Patients with a damaged frontal lobe like Jim perform so poorly in the Gambling Task because they lack the restraining influence exerted by a network of brain structures that include the frontal lobes and various components of the limbic system. As a result of these abnormalities, anxiety is conspicuously absent. The resulting personality deficits show up both on psychological testing and in real life: Jim's failure during the Gambling Task to cease selecting from the fixed decks and his lifelong adjustment problems.

Research carried out on Jim and others like him supports the view that anxiety isn't a liability, but rather a necessary component for normal psychological functioning. People like Jim and others with frontal lobe abnormalities can't experience anxiety, and as a consequence they fail to make decisions that are in their best long-term interest.

Anxiety, it turns out, is an important—indeed, essential—emotional component of our personality that is best thought of as existing along a continuum. At low levels, our anxiety provides a useful constraint on our impulsive, often self-destructive actions. But as the anxiety level climbs upward and reaches sufficient intensity, we experience a kind of psychological paralysis. Think of anxiety as similar to arousal: When we are drowsy, we can hardly function at all; additional alertness activates us into the range of optimal performance; but alertness beyond this point (as a result of stimulant drugs, for instance) leads to overactivation, an ensuing loss of concentration, and a worsening of performance.

At this point, let me cautiously advance a speculation about anxiety considered from the point of view of the relationship between the frontal lobes and the amygdala. During the evolution of our brain, the massive growth of the prefrontal cortex resulted in an increase in back-and-forth traffic between that area and the amygdala. Eventually, the prefrontal cortex specialized in processing the details of our experiences with the amygdala, setting the emotional tone for the experiences. In time, these two areas came into balance.

Over the last several hundred years, this balance has been disrupted thanks to the flood of information flowing into the brain. It's estimated that the Sunday edition of many major newspapers contains more information than people in earlier times processed over the span of several years. The amount of information available to us has increased to the point that the amygdala can no longer handle the traffic shunted to it from the frontal cortex, especially when the information involves matters that are inherently anxiety inducing. The situation is similar to what would happen if a major television production company fed dozens, perhaps even hundreds, of new programs to a community-broadcasting channel capable of airing only a small percentage of the feeds. The community channel simply wouldn't be capable of processing all of the material; any attempts to do so would only result in serious disruptions in transmission.

Think of the amygdala as that community station and the frontal lobes as the major production company. In response to the flood of too much information from the prefrontal cortex, the amygdala begins to fire excessively or under inappropriate circumstances. Anxiety is the end result of this overactivity. Our best response is to strive for a proper balance between the amygdala and the frontal lobes, between too much anxiety and not enough.

## SUGGESTIONS FOR TRANSFORMING YOUR ANXIETY

*Recruit the Help of Your Prefrontal Lobes During Panic Attacks or Especially Severe Episodes of Anxiety*

Since the frontal lobes serve as a check on the limbic system, enhancing frontal lobe functioning can help control your anxiety. Helpful here are exercises like those described in my earlier book *Mozart's Brain and the Fighter Pilot*. Step back from the experience and remind yourself that you've survived such attacks in the past. Years ago, psychoanalysts referred to this as recruiting the "observing ego," by which they meant observing one's own reactions as if watching the reactions of someone else. Although this technique often works, it may not prove successful if your attack is marked by autonomic nervous system (ANS) accompaniments such as an irregular heartbeat, alterations in your breathing, sweating, a feeling of chest constriction, and the like. At such times, you should take the following step.

### Exercise the Brain-Body Connection (BBC)

Thoughts influence bodily reactions and vice versa. During episodes of anxiety, take a few deep breaths while sitting with your eyes closed. Clear your mind so that you're not mentally occupied with anything in particular. Consciously relax the muscles in your neck and shoulders. Then think about any one of the five most wonderful things that ever happened to you in your life, or one of the five accomplishments in your life that you were able to bring about because you were in complete control of your emotions. While engaged in this exercise, your anxiety should greatly lessen or even disappear. What's best about the BBC is that you can work with it starting either with the brain or the body. For instance, when practical considerations preclude your closing your eyes or otherwise modifying your bodily responses, you can always withdraw into yourself and occupy your mind with those pleasant images, primarily the products of your cerebral hemisphere. In this way, you will positively affect the emotional circuitry of the limbic system and thereby influence the hypothalamus, which in turn influences the autonomic nervous system. It's the ANS that mediates such bodily processes as blood pressure, breathing, and muscle tension. Your goal is to help bring your bodily responses under the control of the parasympathetic ("relax and smell the flowers") rather than the sympathetic ("fight or flight") division of the ANS. If you are successful, you'll notice a decrease in your anxiety.

What you don't want to do is attempt to relieve anxiety in ways that will lead to a rebound in anxiety. For instance, never use nicotine as an anxiety reliever. If you don't smoke, don't consider starting; if you do smoke, try to quit. Not that smoking can't temporarily help you feel less anxious; it can (ask any regular smoker). But that anxiety relief comes at a price: A few minutes after the last cigarette, you will start craving another one. "Nicotine is quite effective in reducing separation distress in animals," wrote neuroscientist Jaak Panksepp. "That's the explanation for why the desire for cigarettes among regular smokers is so strong. Nicotine withdrawal heightens feelings of separation-type distress while smoking alleviates such feelings. As a result, nicotine habits are so hard to break: when deprived of cigarettes, the smoker experiences an increase in the intensity of those separation distress-like feelings."

# 7

# Over the Edge
*Anxiety Disorders*

CHIEVING THE PROPER BALANCE between too much and too little anxiety isn't necessarily easy, as I first observed during my medical school days. Several of my fellow students took various approaches to controlling the anxiety aroused in them by the academic challenges they were facing. Although the fail-out rate in our school wasn't high (perhaps one or two per year in a class of less than two hundred), many of the students weren't reassured by such statistics and suffered agonizing anxiety that they would be one of the failures. Their various approaches to dispelling their anxiety included, as I learned later, the key elements of several anxiety "disorders."

One student adopted a bizarre variation on the widespread use among the students of using Magic Marker pencils as a study aid. But instead of employing a brightly colored pencil to underline important sentences in his textbooks, he favored a black laundry marker to obliterate sentences that he considered unimportant. The pages of his textbooks thus consisted of selected sentences displayed against a background of thick wavy black lines.

Another student bound his anxiety by compulsively adopting a favorite seat in the lecture hall. In order to guarantee that another student didn't take his seat, he would arrive half an hour before the scheduled lecture. But on one occasion, heavy traffic delayed his arrival, and on entering the hall he found his seat occupied. I watched as he asked the inadvertent interloper to please move to another seat. Somewhat puzzled at such a strange request, the other student nonetheless complied, much to the obvious relief of his classmate, who promptly snuggled into the comfort provided by his usual cocoon.

Another student we nicknamed the Traveler suffered such severe anxiety while studying alone in his rented room that he gave up the room and moved his belongings into his car. At night, he drove to the various hospitals affiliated with the medical school in search of an empty bed in one of the on-call rooms set aside that night for doctors assigned to duties at the hospital. If a bed wasn't available at one hospital, he "traveled" to another. Always anxious and easily upset about minor matters, the Traveler was often the victim of cruel jokes.

Not all of those living lives organized around anxiety are as obvious as the Traveler. Some of the indicators of anxiety are quite subtle and, since anxiety is an internal state, may be undetectable to the casual observer. As an aid in recognizing anxiety in others, concentrate on what makes you anxious and how you respond. I've found from personal experience that you can learn a lot about yourself from such a self-evaluation.

Several years ago, I began taking note of situations that made me anxious. For instance, I do a lot of public speaking, sometimes to audiences of several hundred people. And while I'm always more alert and focused just prior to going out on the stage, I'm not really anxious. In fact, I enjoy public speaking. In contrast, I become mildly anxious whenever I have to speak up in front of a small group of people, such as while placing an order in a store or giving my name for placement on the seating list at a crowded restaurant. What could be the reason for this curious dichotomy? If I can speak comfortably to hundreds of people, what's the big deal about merely giving my name or placing an order in front of a few people?

By focusing on my anxious feelings in these different situations, I finally solved what had long seemed an inexplicable problem. In the public-speaking situation, my role and identity are firmly fixed: I'm the brain expert, the lecturer, the author, and so on. But in the store and restaurant situations, my identity isn't fixed: I'm at once both somebody and nobody, only one member of an anonymous crowd.

Try to think of situations that make you anxious. Don't randomly theorize about why you're anxious in these situations; simply identify them. Then sit at a word processor and quickly write down what you recall about your experiences and thoughts during those occasions. You'll find that an understanding of the source of your anxiety is more likely to result from analyzing your thoughts and feelings from those earlier anxiety episodes than from abstract theorizing. In my case, I discovered that my anxiety in small public gatherings centered on such concerns as speaking loudly enough to be heard but not loudly enough to attract general attention. My appearance also played a part; specifically, I was anxious about whether I was appropriately dressed for the circumstances—neither too dressed up nor too casual. These insights helped me to recognize that my anxiety arose in the absence of a clearly defined role or identity when among small groups of strangers.

Another effective means of understanding and managing one's personal level of anxiety is to learn about the various anxiety disorders that result when anxiety becomes excessive or uncontrollable. In chapter 2, we encountered literary examples of anxiety in the narrator of "The Tell-Tale Heart" and Mr. Snagsby in *Bleak House*. In this chapter, I'm going to introduce you to real persons afflicted with the most common anxiety disorders, most of them drawn from my own private practice of neuropsychiatry. Although the frequency or intensity of their anxiety may be greater, their basic experience isn't that much different from those of us who, suffering lesser degrees of anxiety, continue to function, albeit perhaps less efficiently than our nonanxious neighbors.

Let's start with the most severe form of anxiety disorder and move down from there to anxiety disorders that are barely distinguishable from perfectly normal experiences.

## POST-TRAUMATIC STRESS

"I know Joan is dead, yet I see her as clearly as I see you right now." The speaker is Wendell, a tall African-American man of forty. He's telling me about a thirty-three-year-old homeless woman simply known as "Joan" who committed suicide three months ago by kneeling on the tracks at a Metro station near the Pentagon. The train operated by Wendell crushed her to death as it entered the station. At the fatal moment, "she looked right at me," Wendell has told me on several occasions during our previous meetings.

Seconds after the impact, Wendell leaped from the cab. He remembers seeing "blood all along the side of the front car." He telephoned the central dispatching station with the curt message "Purple; purple"—code words used by operators to signal that someone has leaped in front of a train. Minutes later, while responding to questions posed by the transit police, Wendell remembers turning distractedly toward the cleanup crew busily engaged in the grisly process of removing Joan's body from the tracks and washing her blood from the train.

That evening, Wendell slept poorly, often awakened by dreams of the woman's face looking up at him from the tracks. And late the next morning, while driving in traffic, he suddenly experienced a fear that a pedestrian might bolt from the pavement and leap in front of his car. Frightened, Wendell immediately returned home; he resolved to limit his future driving to no farther than a few blocks from his neighborhood.

A day later, Wendell experienced shortness of breath and chest discomfort at the thought of returning to the trains or even entering the subway. His travel options now severely limited, Wendell remains largely in his house, fearful of the darkness at night and of the loneliness that comes on after his girlfriend leaves for work in the morning.

Usually an avid reader, Wendell can no longer finish a paragraph or sometimes even a sentence before his mind drifts off and he starts thinking again of "the accident." He keeps the TV on during the day, even when he isn't watching it—"just to have some sound in the background." Ordinarily calm and deliberate, Wendell now tends to startle and overreact to everyday sounds. On one occasion, he became convinced that someone was walking along the hallway outside his bedroom. When he cautiously and fearfully opened the door, the culprit turned out to be his cat. But most distressing to Wendell are the "visitations."

"I can look away from the TV and she's there in the room, staring at me from the tracks with that same expression. 'You're going to pay for this,' she says. I know in my mind that it can't be her because she's dead. Yet there she is!"

One morning, Wendell heard on TV about another suicide in the subway. Immediately he "knew" that Joan was "somehow involved." "She walked into the living room where I was watching TV and said, 'I pushed that person in front of the train and there will be others.'"

The next evening, while sitting in the passenger seat of a car driven by a friend, Wendell thought that he recognized someone he knew standing by the side of the road. "I asked my friend to stop and back up. When he did, nobody was there. I told him that I was certain I had seen someone there moments before and that I believed that person was Joan. I told him she had disappeared when she saw our car backing toward her. At that point, my friend became kind of agitated and said I should go back to your office and talk to you about it. I think my friend thinks that I'm crazy. Maybe I am."

When I first spoke with Wendell—especially when he told me about the visitations—I initially concluded that he was psychotic: totally out of touch with reality. His visitations conformed to a combination of visual and auditory hallucinations coupled with the delusional belief that Joan was still alive. But Wendell doesn't come across as the typical psychotic. He's friendly, likable, makes good eye contact, and speaks sensibly on just about any subject I can bring up—everything, that is, except Joan. When discussing her, he drifts between reality and a world of fantasy and hallucination.

"The other morning I woke up with a bruise on my arm. Then I remembered that during the night Joan had come and twisted me on the arm in order to hurt me."

"Don't you think it's more likely you simply had a dream?" I asked.

"One minute I think I might have accidentally hit my arm during the night and then had a dream about her injuring it, but the next moment I think it really happened. I can remember seeing her and hearing her saying, 'I'm going to hurt you.'"

As this exchange illustrates, Wendell's brain is operating according to a kind of double bookkeeping system. He recognizes the unreality of what he reports, and yet it seems real enough to arouse intense anxiety. At night, he keeps the lights on in his bedroom and listens to background sounds that one moment seem explainable by "the house settling." The next moment, he's convinced the sound indicates the presence of "someone in the house." This alteration between reasonable perceptions and distortions create anger and shame: "I have never been a man who was afraid of things, but now I am scared of everything."

In response to his disturbing experiences, Wendell isolates himself. His girlfriend complains that he's "distant," "unavailable," "irritable," given to "flying off the handle" over trivial matters, and no longer interested in sex. "I just don't want to see anybody," Wendell told me. "They don't believe me when I talk to them about Joan; they just want to tell me all this is impossible, or they want to argue with me about her. I'm just better off staying in my house by myself."

Wendell's diagnosis is post-traumatic stress disorder (PTSD).

IN 1980, THE AMERICAN PSYCHIATRIC ASSOCIATION first recognized PTSD as an anxiety disorder and included it in the third edition of its *Diagnostic and Statistical Manual of Mental Disorders* (DSM-III). The diagnosis required exposure to a traumatic "stressor that would evoke significant symptoms of distress in almost everyone" and "that is generally outside the range of usual human experience." Wendell's experience certainly meets both criteria.

Three clusters of PTSD symptoms resulted from Wendell's trauma. The first cluster consists of intrusive thoughts and dreams about the trauma (Wendell's "visions" of Joan both while asleep and awake) and sudden, acutely disabling "flashbacks" that the trauma is happening again (Wendell's experience of "seeing" Joan looking up at him from the tracks moments before her death).

The second cluster involves numbing: feelings of distance and estrangement from others (Wendell's unwillingness to leave home or socialize), blunted emotions, and loss of interest in formerly enjoyable activities (his distancing himself from his girlfriend).

The third cluster comprises a mix of symptoms including an enhanced startle response, hyperalertness for potential threats, sleep problems, memory and concentration problems, and avoidance of distressing reminders of the trauma and guilt.

In essence, PTSD sufferers like Wendell remain perpetually anxious because they keep reliving the traumatic incident in thoughts, feelings, images, or actions. Think of PTSD as an intense anxiety disorder resulting from unbidden and unwelcome memories of trauma. Further, these memories aren't experienced as a whole but as fragments: isolated sights, sounds, smells, and bodily sensations. Intense waves of distress accompany these memory fragments. The sight of a woman who looks even remotely like Joan induces in Wendell an acute episode of anxiety that on occasion terminates in a panic attack. As a defense against such experiences, Wendell often "spaces out," appears to be out of contact, unresponsive to the people and events around him. On other occasions, he does or says things he can't remember later. Not surprisingly, Wendell feels "out of control" and fears that he "may be going crazy."

In order to keep his anxiety at bay, Wendell organizes his life around avoiding anything that reminds him of his experience. He shuns anyone who raises doubts about Joan's existence. To keep his mind "occupied," Wendell forces himself into repetitive, monotonous activities like raking and reraking the leaves on his lawn. But despite these attempts to lessen his anxiety, recurring feelings of helplessness, foreboding, and tension lead to emotional overreactions, exaggerated startle responses, irritability, and temper outbursts. In response, Wendell's friends and even intimates shy away, thus imprisoning him even more firmly in his anxiety. And since all of his thoughts are threatening and ominous, Wendell tries to force them from his consciousness. Instead of focusing on the moment and thinking of the future, Wendell expends his mental energy on *not* thinking about the past. As a result, he gains no pleasure from current experiences and just wants to "be left alone."

When talking to somebody like Wendell, it's easy to fall into the trap of thinking, as psychiatrists did for many years, that everything is "psychological." According to this view, people who experience a traumatic event are expected to "get over it" and "move on" without any resulting impairment. People who don't comply with these expectations are considered to be exaggerating their difficulties.

Today, scientists know that PTSD isn't simply an exaggerated psychological response to trauma. Specific physical features distinguish the anxiety of PTSD from a normal response to trauma. Initially, the traumatic event causes the hypothalamus at the base of the brain and the amygdala to activate both the startle and the fight-or-flight responses, which involve principally the sympathetic nervous system, the hypothalamus, and the adrenal glands sitting atop the kidneys. For reasons that aren't entirely clear, the adrenals of a person who goes on to develop PTSD usually put out decreased amounts of cortisol, a chemical that enhances the body's ability to handle stress. Decreased cortisol translates into an impaired stress response. Further, that person responds with greater-than-normal activation of his or her sympathetic nervous system. According to experts on PTSD, this combination of lower-than-normal cortisol levels accompanied by greater-than-normal sympathetic nervous system activity may even constitute a biologic risk factor for PTSD.

"Greater sympathetic nervous system arousal leads to an increase in the concentration of adrenaline in the blood," according to Rachel Yehuda, director of the Traumatic Stress Studies Program at the Mount Sinai School of Medicine in New York. "This in turn leads to a greater level of arousal, a state that favors the formation of vivid memories. If a sufficient level of cortisol doesn't offset this increase in adrenaline, the arousal tends to be prolonged. The end result is enhanced memory formation. Later under conditions of increased distress, the level of adrenaline rises once again, leading to a replay of the vivid memory."

Additional insights into the physical basis for PTSD come from modern imaging technologies, which reveal several brain abnormalities.

In one study, combat veterans with PTSD tape-recorded narratives of their traumatic battlefield experiences. When the veterans later listened to this "script-driven imagery" of the traumatic events, the blood flow to their left amygdala increased. No increase occurred in veterans with wartime experience who hadn't developed PTSD. Further, increases in heart rate and other indicators of increased sympathetic nervous system activity accompanied the increase in blood flow to the amygdala.

In another study, PTSD combat veterans participated in what's known as a masked-face experiment, which is based on subliminal perception. The subject stares at a computer screen on which the experimenter displays a picture of a fearful face. The picture appears very briefly, about $^{33}/_{1,000}$ of a second—much too short a time for conscious perception of the face. A second picture of a face showing no emotion immediately follows this. This neutral face remains on the screen five times longer than the fearful face—$^{167}/_{1,000}$ of a second, to be precise. Typically, the subjects consciously perceive the neutral face but fail to report having seen the fearful face. But despite their inability to recall seeing the fearful face, PTSD veterans show an exaggerated amygdala response on fMRI testing, a response that varies directly with the severity of their PTSD symptoms.

The masked-face experiment provides a practical demonstration of Joseph LeDoux's "high road/low road" distinction, described in chapter 4. Because the picture of the fearful face remained on the screen for so short a time, the processing of that image never made it to the cortex for conscious elaboration but remained instead confined to the low road (the amygdala and the thalamus) located deep below the cerebral hemisphere. As a consequence, the PTSD sufferer hadn't a clue why he or she was feeling upset while simply looking at a series of seemingly unremarkable faces.

In contrast to the increased activation of the amygdala, other areas of the brain show *reduced* activation in PTSD. Prominent here is the anterior cingulate, the limbic structure mentioned in the previous chapter that is involved in memory, emotion, and selective attention. As mentioned earlier, the anterior cingulate springs into action whenever concentration is called for. For instance, quickly read aloud or write down the number of repetitions of each word in the list on the next page. Ignore the meaning of the words; simply count and say aloud the number of times each word appears in each vertical column.

| Three | Five | Six | Three | Four | Four | Two | Two |
|-------|------|-----|-------|------|------|-----|-----|
| Three | Five | Six | Three | Four | Four | Two | Two |
| Three | Five | Six | Three | Four | Four |     | Two |
|       | Five | Six | Three | Four | Four |     |     |
|       |      | Six |       |      | Four |     |     |

Note how much easier it is to read the columns in which the word for the numeral corresponds to the number of repetitions ("three" written three times rather than four times). That's because your brain has learned over many years to give precedence to the meaning of a word rather than the number of repetitions. Thus, if the number 4 is written out five times, your brain undergoes a slight but measurable lag before coming up with the correct answer (five). It's the anterior cingulate that helps us to ignore the meaning of the word and concentrate on the number of times the word appears.

A similar conflict occurs if color words are printed out in mismatching inks (i.e., *red* written in green ink, or vice versa). Try it for yourself using colored pencils. Write out twenty or thirty color words (you can use the same color word more than once) with the pencil of the corresponding color and repeat it with another list containing mismatches of word and colored pencil. Then time yourself as you name as rapidly as possible the color the word is printed in rather than the color name. You'll experience greater difficulty when the word name and pencil color don't match. Enhanced activation of the anterior cingulate occurs when you attempt to make that distinction.

Emotions can be introduced into the testing by the use of emotionally charged words instead of number words (e.g., *murder* written four times). And it's on this test that PTSD sufferers perform at their worst. While veterans with PTSD perform like everybody else on tests of color and number mismatches, they show an altered activation of their anterior cingulate when reading trauma-related terms (e.g., *body bag*).

Finally, people with PTSD have smaller hippocampal volume, secondary to the death of hippocampal neurons thought to be caused by increased exposure to elevated stress hormone levels. And since the hippocampus serves as the initial entry point for the formation of new memories, PTSD sufferers typically complain of memory problems and show severe memory deficits on testing.

In short, PTSD involves major disturbances in brain organization and functioning that may sometimes, as with Wendell, lie on the uneasy border separating sanity from insanity.

It's been six months now since I first encountered Wendell. He has returned to work, but not on the trains. Instead, he's resumed a previous job driving a bus. He no longer experiences flashbacks or dreams of Joan. If you had the opportunity to talk to him now, it's unlikely that you would suspect that this relaxed and confident man once suffered from PTSD. What led to his improvement?

Few subjects are as controversial as the treatment of PTSD. If you talk to a psychoanalyst or psychotherapist, he or she will tell you that a cure requires that the PTSD sufferer relive the traumatic experience in order to master it. Commonly employed treatments may include script-driven imagery, mentioned earlier, which involves the PTSD sufferer recording an audio account of the traumatic experience. Later, during treatment sessions, the patient listens to the recording and relives the traumatic experience. I've observed such sessions and can vividly recall the crying, shouting, and other emotional outbursts that typify what can happen when a PTSD sufferer is forced to relive events he or she has made great efforts to avoid thinking about. While such an approach to PTSD may help in the hands of some doctors, I tried something different with Wendell.

Wendell's greatest fear was that he was "going crazy." The flashbacks, sleeplessness, chronic tension, and difficulty concentrating terrified him. If his terror continued to escalate, I feared that he might seek relief by trying to end his life. My first tasks therefore were to calm him down and help him to make sense of his confused and frightening experiences.

"Joan is responsible for what's happening to me," Wendell mentioned on several occasions during our discussions. In order to combat Wendell's delusions about the continued influence of Joan on his life, I gave him a drug known as Olanzapine. Since it is a tranquilizing drug suitable for extreme—even psychotic—levels of anxiety, it could be counted on to calm him by cutting back on his heightened arousal: the jitteriness, hyperalertness, startle reactions, and sleeplessness. But even when ordering the drug, I knew better than to expect a sudden and complete disappearance of all of his symptoms. Wendell wasn't about to be convinced of the impossibility of his beliefs simply as a result of taking a pill. That's because delusions are notoriously resistant to medications. But despite the failure of the medication to totally banish the symptoms of psychosis, the antipsychotic power of the drug lessened the delusions and hallucinations.

Along with the medication, I played the role of teacher and explained to Wendell in everyday language the symptoms of

PTSD. While he was vaguely familiar with PTSD, most of his knowledge came from movies and TV docudramas. This was of little help to him, since he couldn't relate his current feelings to fictional re-creations of the disorder. Instead, he had taken his anxious fears at face value and concluded that Joan was haunting him—and, further, would continue to do so for the rest of his life. In response to this conviction, he had begun organizing his life around the overriding goal of avoiding anything that reminded him of what he referred to as "the accident."

Thus, my first challenge was to gently call into question Wendell's belief that Joan was still influencing events in his life. Another goal was to teach Wendell ways to regain a sense of control over what was happening to him. And that control could only come about when he could once again trust his own perceptions and feelings. As a first step, he had to learn to identify and accept his anxiety without giving in to panic. In short, Wendell's treatment involved the initial prescription of a tranquilizing and antipsychotic drug followed by active efforts to provide alternatives to his delusions about what was happening to him.

Other treatment approaches are based on our emerging understanding of the brain processes that go awry in PTSD. For instance, one theory about flashbacks involves a feedback loop. Each reexperience of the trauma results in the release of stress hormones, principally epinephrine or its cousin, norepinephrine (known collectively as adrenergic hormones because of their chemical structure). These hormones, in turn, feed back via the bloodstream to the brain, where they strengthen the memory of the trauma and thereby increase the likelihood that the traumatized person will reexperience the trauma again and again.

At this point, no one can prove that PTSD is based on a feedback loop. Nevertheless, the theory suggests a strategy to prevent PTSD: suppress the adrenergic activity with drugs either prior to or shortly after a traumatic experience. If the output of adrenergic hormones can be suppressed or greatly reduced, a person's memory for the traumatic aspects of the experience should be weakened, making flashbacks less likely to occur.

In a test of this approach, a group of auto accident survivors received the drug propranolol a short time after their accident; a control group with similar experiences wasn't given the drug. Propranolol works by blocking the adrenergic receptors both in the brain and in the peripheral nervous system.

Both groups were later tested for signs of arousal (increases in heart rate and breathing, sweating, etc.) while listening to audiotapes (made by the survivors themselves) describing the accident. The propranolol-treated survivors showed fewer signs of arousal in response to the taped narratives, compared with those listening to taped narratives who hadn't received the drug.

Futuristic uses of propranolol-like drugs could include their administration to ambulance workers headed out to the scene of a disaster; the survivors of the disaster could also be given the drug at the scene as part of their initial treatment. The drug wouldn't erase memories of the disaster, but it would effectively diminish the emotionally traumatic aspects of those memories. As a further development, there are some suggestions that even if the drug isn't immediately available, it might be effective later to alter already established memories. What could be the basis for this ability of a drug to alter something as permanent as a memory?

First of all, memories aren't nearly as permanent as many of us have been led to believe. Over the years, we suffer losses in the richness and specificity of our memories. For instance, think back to your high school graduation. You remember less about it now than you did during those first few years after it occurred. With each passing year, you remember less (unless you're regularly reminiscing about your graduation for some reason). This normal "forgetting" is mirrored at the molecular level by alterations in the efficiency of protein synthesis.

Each time you think of your graduation, you consolidate that memory via synthesis of specific proteins in the memory encoding areas within your brain. And if that protein synthesis is interfered with, the memory isn't successfully reconsolidated. The title of a pivotal scientific paper on the subject provides a capsule summary of the process: "Fear Memories Require Protein Synthesis in the Amygdala for Reconsolidation After Retrieval." The researchers found that rats that have been fear conditioned lose their fear if the experimenter takes measures to interfere with protein synthesis prior to the consolidation of the fear memory. Researchers are presently exploring the possibility that a similar interference with protein synthesis may prevent the reconsolidation of the emotional components of a person's memory for a traumatic event.

But let's get back to Wendell. Over the space of four months, he gradually recovered in response to medications and some discussion about the factors responsible for his anxiety and PTSD. Today he looks back on the experience with puzzlement: "I can't understand how I could have felt that way and said those things. I guess I must have been pretty upset and didn't know how to respond to the anxiety I was feeling."

Wendell's experience illustrates the delicate balance between health and illness that exists within all of us. Prior to the incident, Wendell wasn't particularly anxious and lived a normal life. Yet his whole life changed as a result of witnessing something that took place in a matter of seconds. This, of course, leads to the question you may be asking yourself at the moment: If such a severe and devastating anxiety response can occur in a person like Wendell, who had no previous signs of psychological maladjustment, could a similar response occur with *me?*

From my observations of Wendell and others with PTSD, I've come to the conclusion that given a sufficiently traumatic event, anyone can develop PTSD. Fortunately, most of us will be spared traumatic experiences of such severity. But it's a sobering thought, nonetheless, that we are all potentially at risk of developing PTSD. Further, if the anxiety becomes great enough, any of us might, as Wendell did, escape into a fantasy world. Certainly, Wendell's PTSD developed after he underwent an experience that anyone would find stressful. But other people appear to develop the condition following far less traumatic experiences.

Experts disagree about the existence of milder forms of PTSD. According to the DSM-IV definition, the causative event must involve "actual or threatened death or serious injury, or a threat to the physical integrity of self or others." Yet some people with the diagnosis of PTSD attribute their condition to experiences that many people wouldn't consider all *that* unsettling (being abruptly fired from a job or experiencing sexual harassment or discrimination). Granted, such experiences can inflict emotional trauma, but it's a stretch, it seems to me, to equate them with the threat of death or serious bodily injury. Nonetheless, the patients complain of many of the classical symptoms of PTSD. What's more, it doesn't take more than a few minutes of talking with them to become convinced of the depth and intensity of their suffering. Despite this, I believe that to label such people as suffering from PTSD trivializes the disorder. Moreover, events that now qualify as precipitants for PTSD would be considered only minor irritants in an earlier era.

"One unintended consequence of peace and prosperity is a liberalized definition of what counts as a traumatic stressor," said psychologist Richard J. McNally. "The threshold for classifying an experience as traumatic is lower when times are good."

Among the PTSD-inducing stressors cited by McNally are age discrimination, living within a few miles of an explosion (even though one is unaware at the time that it had happened), being kissed in public, seeing the movie *The Exorcist*, and (trust me, this was an actual case) accidentally killing a group of frogs with a lawn mower.

"Referring to such stressful but noncatastrophic events as traumatic stressors produces a kind of conceptual bracket creep whereby increasingly trivial events are awarded causal significance as triggering PTSD. If full-blown PTSD does emerge in the wake of these stressors, then this seems to point to personal vulnerability factors as the causal culprit, not to the event itself," said McNally. In other words, one person's harrowing experience may be the provoking agent leading to another person's PTSD.

While it's generally true that the greater the exposure to trauma, the greater the likelihood of PTSD, there are many recorded exceptions to this rule. Among auto accident survivors, for instance, the seriousness of their injuries fails to predict PTSD symptom severity. "There is no straightforward relationship between the severity of the trauma and the severity of PTSD," McNally concluded.

And there are additional recent changes in our ideas about PTSD. Traditionally, PTSD sufferers were often incapacitated by their condition and essentially unemployable. But many people who fit the criteria for PTSD today remain at their jobs. Typical of this group is Paul Gonzales, a supervisor in the Defense Intelligence Agency's comptroller's office at the Pentagon. During the September 11, 2001, terrorist attack on the Pentagon, seven of the eighteen workers in Gonzales's office died and five were hospitalized. Gonzales is only too aware that he could have been among the casualties. "For a lot of us, a matter of seconds determined whether we lived or died," said Gonzales sixteen months after the attack. In an interview with the *Washington Post* in early February 2003, Gonzales spoke of being "jumpy" and hearing noises at work during the day that "scare the heck out of you." Gonzales continues to work despite experiencing unmistakable symptoms of PTSD.

If you didn't know about Paul Gonzales's trauma, you could easily attribute his tenseness, inability to relax, heightened restlessness, and feelings of being keyed up and on edge to generalized anxiety disorder (GAD). But prior to the terrorist attacks, Gonzales showed none of the signs of GAD. His experience suggests that milder forms of PTSD may develop in people with little or no previous anxiety. What's worse, no one can predict in the individual case those who will go on to develop PTSD after a traumatic experience. Further, it's likely that less-disabling forms of PTSD will become more common among us as a result of the increasing terrorist threat levels that we are experiencing. And, as I mentioned earlier, I believe we are all at risk, given sufficient trauma exposure, of developing at least some of the symptoms of PTSD. It's an anxiety-arousing thought in itself that in response to trauma—varying from person to person on the basis of poorly understood vulnerability factors—any one of us could be in the same situation as Paul Gonzales.

Indeed, picture in your mind an anxiety continuum with Wendell at one end, Paul Gonzales somewhere near the middle, and the rest of us, who cope with our personal degrees of anxiety as best we can, toward the other end.

## PANIC

By adulthood, we learn that while we can retain only limited control over circumstances and events around us, we remain in control of our own responses. But we also learn early in life that under certain circumstances we can lose that sense of inner control—with devastating consequences for our peace of mind.

One of the most distressing experiences of loss of control occurs during a panic attack. Arising without warning and reaching their full intensity within ten minutes, panic attacks can arouse in their victims fears of impending death. Indeed, panic sufferers often cry out in desperation that they can't breathe; that their heart is beating in an irregular rhythm or is about to stop altogether; that their throat is closing over; that their hands and feet are becoming numb and paralyzed. Added to these physical sensations are fear of imminent death, feelings of unreality, and, most disturbing of all, the conviction that one is losing mental control—in essence, "going crazy."

Rather than occurring separately from other anxiety disorders, panic episodes usually erupt as acute flare-ups emanating from a smoldering underlying anxiety disorder such as GAD, obsessive compulsion, or social phobia.

What follows is the experience in her own words of Lisa, a twenty-five-year-old second-year medical student who a year earlier had first consulted me because of anxiety and sleeplessness that started around the time of her final exams. Since at that time she was suffering from an anxiety disorder, I had placed her on a medication that increased the levels of serotonin, an important neurotransmitter (chemical messenger) within the brain. In response, her anxiety had subsided, and she did well on the exams. Indeed, she did so well that she soon forgot about how upset she had been previously and neglected to keep her follow-up appointment. She had returned to my office several weeks before her midterms because of a recurrence of her symptoms. Here in her words is an account of how generalized anxiety can turn into a panic attack of devastating and incapacitating ferocity.

"When I returned home last summer during a break, my doctor said she thought I no longer needed the medicine you prescribed last semester and told me to discontinue it," Lisa begins. I note that her eyes are red and her makeup smeared from crying. "And for a while it seemed like she was right. After stopping the medicine, I didn't notice any difference, but apparently my family and friends did. They said I became cranky and hard to get along with. But nothing really serious happened until three weeks ago. Just like last year, I started to fall apart as I got closer to my exams. First, I started having trouble getting to sleep. Then, two days before the first exam, I experienced episodes when I couldn't get my breath, my heart started racing, and I became convinced I was going insane.

"Somehow I got through the first exam, but on the nights prior to the other exams I couldn't sleep even for a minute. Even more terrifying, my mind started to engage in crazy thought patterns. For example, an hour or so after the second exam, I became anxious that somehow I hadn't turned in my test sheet, even though I distinctly remembered handing it to the monitor and seeing him put it on the desk with the others. I knew it was irrational of me to think I hadn't turned it in, but despite that knowledge I couldn't escape the thought.

"That evening, I called my mother and told her I was going to flunk out of medical school because they would find out that I hadn't turned in my answer sheet. Even as I was telling her this I knew how crazy it must have sounded. 'Of course you handed it in!' my mother assured me. This made me feel better for a while, but then I started to doubt again, and I called my mother back and went through pretty much the same story.

"While talking to my mother the second time, my heart started racing so fast I became convinced that I was having a heart attack. When I told her that, my mother became alarmed and told me to go to the nearest emergency room. The doctors there put me on a heart monitor, and when the results turned out normal they released me with a prescription for tranquilizers. And even though I still have the pills and carry them around with me, I've been afraid to take them—scared I might die of an overdose.

"And now my thinking has ratcheted up to a higher level of anxious irrationality. I can't get away from the thought that I am going to flunk out of school and that I should just leave before the exam grades are back. And that's not the only crazy thought. It doesn't matter what the subject, I take it to the most extreme level. And my thoughts follow a cycle: I'll worry about something irrational, but eventually I'll get myself under control. A few minutes later, a new thought will come and I'll worry about that. The worry can be about almost anything. Did I lock the door? Did I send in my rent check, and if so, will it be received on time? If I'm put out of my apartment for non-payment, where will I live? How can I study if I don't have a room to sleep in?

"In addition to the thoughts, I have these awful feelings. My chest always feels tight. I feel nervous all the time; always expecting something truly awful to occur. I'm so upset by these feelings I don't want to go anywhere near the school in order to reduce the chances that the panic may start over again. For the same reason, I find myself avoiding my apartment. I don't want to be anywhere that reminds me of what I went through.

"The worst panic attacks usually come on at night. I can't take full breaths. I'm dizzy, and my heart starts racing again. The attacks are so frightening that I feel like I'm entering a dark tunnel and that I'm not going to come out into the daylight again. At such times I just want to die. Last night I wanted the panic to stop so much that I felt like hurting myself. Not that I would commit suicide or harm myself directly, but I was willing to knock myself out in order to stop this panic. At one moment, I imagined myself getting into a car accident. I wasn't thinking of deliberately driving head-on into another car, but I almost wanted someone to hit me so that I could get all this to an end. I just feel like I'm coming apart at the seams. I just can't seem to get through the day without having one of these episodes."

Note that in the space of only a few sentences Lisa has described experiences characteristic of four different anxiety disorders. Feeling generally anxious around exam time (GAD), Lisa starts experiencing sudden upsurges in anxiety (panic attacks) as the exams loom closer. She then tortures herself with the thought that she hasn't turned in one of her exams (obsession) and can only find comfort by repeatedly calling her mother to seek reassurance that she had submitted her exam to the monitor (compulsion). As another example of obsessive compulsion, she carries around with her the tranquilizers given to her in the emergency room but fears to take any of the pills because she might somehow overdose on them. Finally, she attempts to shield herself from another panic attack by avoiding her school and apartment (phobic avoidance).

After placing Lisa back on the same medication that had helped her before, I decided to check into what might be available on the Internet to people experiencing panic experiences similar to hers. I logged on to my favorite site, Google, and entered the key word *anxiety*. Within less than a minute of searching, I discovered several online tests for anxiety. The

Anxiety Quiz developed in 1988 by Sir David Goldberg and two collaborators published in the *British Medical Journal* claims to be able to help people determine how their anxiety levels compare with those of other people their age. It consists of a simple yes or no response to the following questions.

Have you felt keyed up or on edge?

Have you been worrying a lot?

Have you been irritable?

Have you had difficulty relaxing?

Have you been sleeping poorly?

Have you had headaches or neckaches?

Have you had any of the following: trembling, tingling, dizzy spells, sweating, diarrhoea, or needing to pass water more than usual?

Have you had difficulty falling asleep?

Take the quiz and check your answers at http://bluepages. anu.edu.au/symptoms_anxquiz.html. My own responses evoked this computerized response: "You scored 6 on the Goldberg anxiety quiz. About 26% of adults report as many or more anxiety symptoms than you and 74% report fewer symptoms. This suggests that you have a moderate number of anxiety symptoms."

If you're up for another quiz, answer the "Are You Anxious?" questions in the "Emotional Health Quizzes" section featured on the website of psychiatrist Harold Bloomfield (www.haroldbloomfield.com).

While such tests provide a rough measure of your anxiety level, they're fairly crude instruments. A person can score normally on the tests and still be anxious. Conversely, a person might score higher than his or her anxiety level warrants.

Another way to learn about your anxiety level would be to consult any one of the hundreds of anxiety-related websites that cater to the anxiety preoccupied. This mix of the authoritative and the kooky even includes the Anxiety Network Bookstore (www.anxietynetwork.com/bookstore.html), which invites the anxiety sufferer to log on to any one of 220 anxiety-related books with titles such as *Abolish Anxiety: Discover Inner Peace in a Stressed-Out World*; *Dancing with Fear: Overcoming Anxiety in a World of Stress and Uncertainty*; *Finding Serenity in the Age of Anxiety*; and *How to Control Your Anxiety Before It Controls You*.

Entering the Anxiety Network Bookstore, I encountered right up front the message "Click here if you are having a panic attack." I mentally placed myself in Lisa's position and clicked "here" in order to find out what help was available. I came up with the following.

## PANIC ATTACK NOTES

These reminder notes have been developed by an editor with more than sixteen years "of experience in dealing with panic attacks." If you are having a panic attack, repeat these reassurances, coping skills, and positive affirmations to yourself **slowly and often**.

*Reassurances:*

- You are in complete control. No harm will come to you at all. You are not in any danger.

- The feelings of comfort will come. The feelings of discomfort will end!
- You have absolutely nothing to fear. These are just harmless feelings and harmless thoughts.
- It's a bluff and a liar. You may think this is different from other panic attacks, but it is not. It's the same old anxiety and it is benign.
- Trust these facts. Do **not** trust your feelings.
- Feelings come and go. You will calm down.

In addition, the editor suggests the following coping skills.

- Slow Down! Force yourself to breathe slowly and deeply.
- Repeatedly stop any negative thought, image or memory.
- Be patient and let time pass. Peace and comfort lie on the other side of anxiety.
- Accept your feelings, don't fight them. Let them go and don't add 2nd fear.
- Focus your attention externally. Tune out your body.
- Do some thought restructuring. How likely are the feared events going to occur? How well can you cope even if they do?

While such injunctions may be helpful to Lisa or others in the throes of a panic attack, they call for an act of faith that, it seems to me, may sometimes be beyond the powers of the panic sufferer. If Lisa is experiencing chest pain and shortness of breath, how can she be *certain* that this time she isn't suffering a real heart attack rather than anxiety-driven chest discomfort? In other words, doesn't she have to wait until the episode is over before definitively labeling it as a panic attack? Certainly, the experience of previous panic attacks doesn't guarantee that this time she isn't encountering a potentially life-threatening illness.

Further, advice such as "Do not trust your feelings" assumes a dichotomy between feelings and thinking that—and this is one of the main themes of this book—doesn't correspond to the reality of our inner experience. Activation of the amygdala and other limbic structures can on occasion result in anxiety powerful enough to overcome any reassurances provided by the frontal lobes.

To the person undergoing a panic attack, the uncontrollable eruptions from the emotional brain seem realer than real. That's why even psychiatrists can have panic attacks and not recognize them. While their frontal lobes may provide the necessary information ("This feels like a panic attack and certainly fits the description given in DSM-IV"), the intensity of their emotions prevents them from putting their experience into perspective.

Figure 3 is an explanatory model of panic disorder. First, a physical symptom leads to misinterpretation ("I have a tiny bit of discomfort in my chest. I haven't felt anything like this before so I must be having a heart attack"). This misinterpretation leads to increased scrutiny and monitoring of how things "feel" in the chest, which leads to anxiety. Additional scrutiny and escalating anxiety eventually results in a confirmation of the belief that one is experiencing a heart attack. Other more likely possibilities for the feeling in the chest (indigestion, muscle sprain, needless worry, etc.) are simply ignored. Eventually, the belief in the "reality" of the heart attack explanation becomes so compelling that a panic attack ensues.

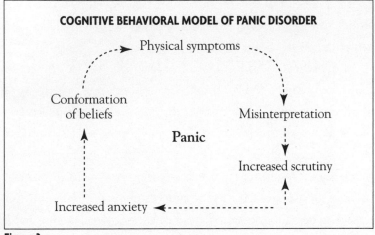

**Figure 3**

While the severity of the first panic attack may vary, one-third of those experiencing their first attack flee to the nearest emergency room. After examinations and tests have established that the feeling of a heart attack was due to something else—perhaps muscle strain after too many hours spent working on some pectoral exercises—the panic attack sufferer returns home greatly relieved. But relief is short-lived. Memory of the mounting anxiety that culminated in the panic attack induces anxiety sensitivity: fear of anxiety-related sensations.

Thanks to anxiety sensitivity, the anxious person's attention shifts to bodily processes that normally operate outside of conscious awareness. This shift in focus from the outside to the inside world perpetuates the cycle shown in Figure 3. Even a slight shortness of breath leads to the fear that one's "throat is closing over"; a minuscule increase in pulse rate foreshadows a deadly heart arrhythmia; light-headedness, sweating, or nausea "prove" that despite the reassurances of the emergency-room doctors, a heart attack is indeed on the way. The cycle continues until the anxiety breaks out into a full-fledged panic attack.

The higher an individual's anxiety sensitivity, the more likely he or she is to interpret even mild anxiety symptoms as alarming, dangerous, or threatening. Included here are many of the experiences reported by Lisa: *physical* anxiety symptoms, such as sensations of shortness of breath, racing heart, lightheadedness, sweating, tremor, choking, chest pain, faintness, or nausea; along with *mental* symptoms, such as feelings of unreality or being detached from oneself, the sudden impulse to flee, fear of dying, or fear of losing control and going "crazy." Such sensations are so uncomfortable that the person given to panic attacks will do almost anything to prevent or escape them. The most common response is to avoid the situation in which the attack first occurred. Agoraphobia (fear of open places, such as markets) is the most common phobia. The term stems from the Greek *agor:* open market or meeting place. On page 151 is a partial list of the phobias that can result from attempts to avoid panic attacks.

Treatment of panic attacks starts with identifying and altering the anxiety-provoking thoughts. The mantra is "Avoid Catastrophic Interpretations"—negative and harmful explanations of moment-to-moment experiences. Anxious people learn that even though they may not have control over what's happening to them at any given moment, there is one thing they can bring under their control: the attitude they choose to take in response to their anxious thoughts. Rather than giving in to the direst interpretations of their experiences, they're

## NINETY-NINE PHOBIAS

AEROPHOBIA: air
ZOOPHOBIA: animals
MELISSOPHOBIA: bees
EROTOPHOBIA: sex
APIPHOBIA: bees
NEPHOPHOBIA: clouds
OCHLOPHOBIA: crowds
HEMATOPHOBIA: blood
AMATHOPHOBIA: dust
AILUROPHOBIA: cats
DROMOPHOBIA: streets
ANGINOPHOBIA: choking
GALEOPHOBIA: cats
PSYCHROPHOBIA: cold
HIPPOPHOBIA: horses
HYGROPHOBIA: moisture
NYCTOPHOBIA: night
SCIAPHOBIA: shadows
EOSOPHOBIA: dawn
BLENNOPHOBIA: slime
ORNITHOPHOBIA: birds
BATHOPHOBIA: depths
MYSOPHOBIA: dirt
DENDROPHOBIA: trees
PEDOPHOBIA: children
PHAGOPHOBIA: eating
COPROPHOBIA: feces
OMMATOPHOBIA: eyes
PHOBOPHOBIA: fear
ANTHOPHOBIA: plants
PYREXIOPHOBIA: fever
OPHRESIOPHOBIA: smell
CHIONOPHOBIA: snow

FEBRIPHOBIA: fever
PYROPHOBIA: fire
ICHTHYOPHOBIA: fish
ANTLOPHOBIA: floods
HOMICHLOPHOBIA: fog
THERMOPHOBIA: heat
SITOPHOBIA: food
HYLOPHOBIA: forests
BATRACHOPHOBIA: frogs
PHASMOPHOBIA: ghosts
PARTHENOPHOBIA: girls
HYALOPHOBIA: glass
CIBOPHOBIA: food
URANOPHOBIA: heaven
STYGIOPHOBIA: hell
OIKOPHOBIA: houses
LYSSOPHOBIA: insanity
CHRONOPHOBIA: time
ACAROPHOBIA: mites
PHOTOPHOBIA: light
DIPSOPHOBIA: drinking
ONEIROPHOBIA: dreams
CHREMATOPHOBIA: money
KONIOPHOBIA: dust
MUSOPHOBIA: mice
GYMNOPHOBIA: nudity
ONOMATOPHOBIA: names
BELONEPHOBIA: needles
CARCINOPHOBIA: cancer
HOMILOPHOBIA: sermons
LALIOPHOBIA: speaking
COITOPHOBIA: coitus
ODONTOPHOBIA: teeth

ALGOPHOBIA: pain
THALASSOPHOBIA: the sea
PENIAPHOBIA: poverty
CREMNOPHOBIA: cliffs
CYNOPHOBIA: dogs
OMBROPHOBIA: rain
POTAMOPHOBIA: rivers
HARPAXOPHOBIA: robbers
ODYNOPHOBIA: pain
HAMARTOPHOBIA: sin
HYPNOPHOBIA: sleep
SIDEROPHOBIA: stars
GERONTOPHOBIA: elders
MYRMECOPHOBIA: ants
HELIOPHOBIA: sunlight
BRONTOPHOBIA: thunder
HODOPHOBIA: travel
PHENGOPHOBIA: daylight
TOCOPHOBIA: childbirth
EMETOPHOBIA: vomiting
ANEMOPHOBIA: wind, drafts
SELENOPHOBIA: the moon
HELMINTHOPHOBIA: worms
GRAPHOPHOBIA: writing
NECROPHOBIA: corpses
RHABDOPHOBIA: magic
PEDICULOPHOBIA: lice
POGONOPHOBIA: beards
PHARMACOPHOBIA: drugs
LIMNOPHOBIA: lakes
ERGASIOPHOBIA: work
THEOPHOBIA: God
BASIPHOBIA: walking

Source: Robert Schleifer, *Grow Your Vocabulary by Learning the Roots of English Words* (New York: Random House, 1995), p. 143.

taught to treat their thoughts as only hypotheses. Thus, the sensations that have led to anxiety (shortness of breath, dizziness) are recognized as transient and harmless. Most important is learning about the disorder so that each attack can be recognized for what it is: an intense, focused eruption of anxiety resulting from misinterpreting one's own thoughts and experiences. In those instances when this cognitive approach isn't sufficient, medications can be employed that influence the neurotransmitter balance in various areas of the brain.

While we will have quite a bit more to say about neurotransmitters in chapter 9, allow me to briefly mention at this point a few words about the relationship between panic attacks and the interplay of three important neurotransmitters.

A researcher in the early 1980s showed that electrically stimulating a collection of noradrenaline-producing cells in the brain stem known as the locus coeruleus produced fear responses in rhesus monkeys. During stimulation, the animals looked and acted as if they were about to be overcome by a predator. The same response occurred if the monkeys were given drugs that increased the firing of the locus coeruleus cells.

Neuroscientists now know that at least two other central neurotransmitters are involved in panic: serotonin and gamma-aminobutyric acid (GABA). Effective treatment can be directed at one or more of these neurotransmitters. That's because when a drug influences the activity of one neurotransmitter, it influences the activity of the other two as well. Think of the mutual interaction of the instruments in a string quartet; only here we're talking about a string trio of noradrenaline, serotonin, and GABA.

Drugs that decrease the firing rate of the locus stop the panic. Included here are those drugs that directly influence the noradrenaline-containing neurons, as well as other agents that act indirectly on those neurons, such as Prozac and other drugs that affect the concentration of serotonin in the synapse, and tranquilizers like Valium that work at the GABA receptor.

For now, it's sufficient to say that the same neurotransmitters serve as key players in the genesis of panic, generalized anxiety disorder (GAD), PTSD, and OCD. This raises an important question: Why do some anxiety-prone people experience chronic feelings of anxiety, while others experience panic attacks and still others seem entirely free of anxiety until they experience an emotionally traumatic experience?

At the moment, no one has a satisfactory answer to that question. Inheritance no doubt plays a major role in how much anxiety we experience and how we manage it. For some of us, the anxiety remains under control, and we aren't afflicted with an anxiety disorder. Others aren't so lucky. They have inherited a tendency to develop dysfunctional levels of anxiety with the specific pattern (OCD, GAD, etc.) depending on presently unknown variables. For Wendell, the incapacity wrought by anxiety was completely disabling. While Lisa's anxiety wasn't quite as disabling, it was sufficiently severe that she experienced great difficulty managing examinations and even at one point considered harming herself.

Let's complete our survey of anxiety disorders by taking up two others.

## SOCIAL ANXIETY DISORDER

We live in a hierarchical society. Although all of us may be equal in the courtroom, no such equality exists in our homes and offices. Fame, income, and social connection are only some of the factors that bestow special status. And our species has developed ingenious ways of bestowing and withholding that status: greeting rituals (who speaks first and with what degree of familiarity), tones of voice, eye contact, and body posture. A downward, averted gaze or blushing serve as appeasement displays signaling submission, while cold sustained eye contact conveys dominance and power over others.

Given this emphasis on status within the social hierarchy, it's no surprise that some people experience persistent, sometimes crippling anxiety about how others are responding to them. Sufferers from social anxiety disorder (SAD), also known as social phobia, avoid social encounters because of the fear that they somehow don't measure up. They're anxious that they may say or do something embarrassing or humiliating; perhaps a casual remark may expose them as bumbling idiots. In addition, people with SAD remain chronically anxious that other people may be making critical remarks about their clothes, the neatness of their office, the neighborhood where they live—almost any criticism could conceivably be directed at them.

The avoidance, anxious anticipation, and distress associated with SAD makes normal life almost impossible. The SAD sufferer is either actively avoiding the threatening situation or, when avoidance is impossible, breaking out into disabling anxiety attacks. Even though many Americans—including many doctors—have never heard of SAD, more than 5 million Americans suffer from it, according to the National Institute of Mental Health.

Most people with SAD develop the condition early in life (age twelve is the average age of onset). The disorder can be either generalized (related to such things as making general conversation or attending parties or other social events) or circumscribed (occurring in response to specific situations, especially "performances" such as public speaking).

Polls have consistently shown that fear of public speaking ranks number one on most people's list of anxiety-provoking social activities. Included among the other most anxiety-provoking experiences are participating in a meeting, talking to "authority" figures, returning items to a store, or introducing oneself to strangers. Individuals with SAD are especially unstrung at the prospect of any of these activities. As with the general public, fear of public speaking heads the list. Indeed, a SAD sufferer will do almost anything to avoid addressing a group, even a small one. Having to say even a few words provokes severe, sometimes overwhelming anxiety.

Like most anxiety disorders, SAD runs in families: Relatives of people with social phobia have a threefold increased risk compared with other people (16 percent versus 5 percent, respectively). And because the condition is so common within the population (about 8 percent of the population), SAD sufferers often escape detection. After all, who isn't a bit shy on occasion? But more is involved than simple shyness. Indeed, the consequences of SAD can extend over a lifetime. For example, a person with the disorder is almost twice as likely to fail a grade or quit school before graduating from high school.

SAD also exerts a powerful influence on the selection of a career. Thus, careers such as law or teaching are less appealing than computer science or other disciplines that involve communication with machines rather than people.

Not surprisingly, most people with SAD are depressed. Because of their reclusiveness and evident discomfort in social situations, other people tend to avoid them, further increasing their isolation, loneliness, and depression. All too frequently, alcohol provides temporary relief. When they're alone, sufferers from social anxiety drink to assuage their sense of isolation; when they're with others, they drink to lessen their anxiety. Indeed, SAD sufferers comprise a sizable proportion of those people regularly encountered at cocktail parties who undergo a dramatic increase in their sociability after a few drinks.

PET scans provide a vivid portrait of what might be taking place in the brain of the SAD sufferer. During public-speaking performances, reluctant, shy speakers show decreased blood flow to the cerebral cortex, along with a corresponding increase in flow to the amygdala. In contrast, PET scans of people who don't suffer from SAD and aren't overly anxious during public speaking show increased cortical blood flow—an appropriate arrangement, since the cortex is called into play when composing and delivering a speech.

Taken together, these findings suggest that individuals with SAD are prone in social situations to activate areas associated with anxiety while decreasing the activation of parts of the brain involved in pleasure, as well as thinking and planning. But as with the chicken-and-egg question, could we be confusing cause and effect? Is the person with SAD anxious because of some defect in the amygdala, or could anxiety stem from a defect in the neurotransmitter dopamine operating elsewhere in the brain, perhaps within the pleasure circuits? At this point, it seems fair to say that neuroscientists haven't discovered specific brain circuitry alterations characteristic of people with SAD.

One of the impediments to achieving a satisfactory physical explanation is the difficulty in reliably distinguishing SAD from shyness, which is exceedingly common. Based on interviews, between 40 percent and 50 percent of young adults describe themselves as shy. In addition, many of us are shy in some situations while perfectly comfortable in others. Thus, if you lump shyness and social anxiety together, at least 40 percent of otherwise normal young people would have to be considered as afflicted with a psychiatric illness!

But despite many similarities, SAD and shyness differ in several ways. While shy people may tense up prior to a business meeting or an important date, they usually manage somehow to muddle through. SAD sufferers, on the other hand, don't muddle well and consequently will go to great lengths to avoid social encounters. If avoidance isn't possible, an intense anxiety reaction may erupt, with symptoms ranging from sweating and light-headedness to an outright panic reaction.

It makes more sense, therefore, to think of shyness and social anxiety as existing along a spectrum. Shyness is common and best not considered a disorder at all. SAD, on the other hand, represents an extreme; just like compulsive hand washing represents an extreme dysfunctional exaggeration of neatness and cleanliness.

Shyness and SAD can also be conceptualized in dimensional terms: Just as depression represents an exaggerated version of the low spirits we all experience on occasion, social phobia represents an exaggerated form of shyness. "Shyness is a trait that, at its extreme, becomes impairing and thereupon assumes the characteristics of social anxiety or phobia," concluded social anxiety specialists Denise Chavira and Murray Stein of the Department of Psychiatry, University of California, San Diego.

Obviously, it would be valuable to come up with a reliable test for detecting shyness early in life. And neuroscientists have recently done just that with the discovery of a brain "signature" for shyness.

In a fascinating functional magnetic resonance imaging (fMRI) study, Carl Schwartz and his colleagues at Harvard discovered that the brains of twenty-year-olds who had been described as "inhibited" or "shy" at the age of two exhibit a different pattern of activation in response to novelty. When the formerly shy child turned adult looks at pictures of unfamiliar people, his or her brain shows higher activity in the amygdala. According to Schwartz, the fMRI findings "show that the footprint of temperamental differences observed when people are younger persist and can be measured when they get older." As a practical application of this research, fMRIs could serve as an early risk indicator. In shy children with this brain "signature," treatment could be started aimed at preventing future anxiety disorders.

Over the next decade, we can expect to see additional imaging research aimed at the early detection and prevention of anxiety at all points along the shyness-SAD spectrum. On the basis of this research, doctors will be able to recognize and treat the disorder in its earliest stages. With luck and the application of some common sense, the incidence of SAD will be greatly reduced without intruding into the lives of shy but otherwise essentially normal and mentally healthy people.

## OBSESSION AND COMPULSION

My former psychiatry teacher, Leon Salzman, one of the world's experts on the psychology of obsessive-compulsive personality prior to the advent of medications to treat the disorder, provided a succinct explanation for OCD: "The primary dynamism in all instances involves attempts to gain control over oneself and one's environment in order to avoid or overcome distressful feelings of helplessness. The greatest anxiety results from worry about losing control by being incompetent, insufficiently informed, or unable to reduce the risks of living."

This need for control can eventually dominate a person's life. In the words of the late British psychiatrist Anthony Storr, "Perhaps the most striking feature of the obsessional temperament is the compulsive need to control both the self and the environment. Disorder and spontaneity must be avoided as far as possible, since both appear threatening and unpredictable."

Of course, attempts to gain some control over life's uncertainties aren't restricted to persons with obsessive-compulsive anxiety disorder. For most of us, a goodly portion of the day is taken up with trying to gain or restore control over things that during our more reflective moments we may recognize aren't actually subject to our control. Typically, this need for control correlates with our stress and fatigue cycles. When stressed, we fantasize about various work and home problems responsible for our stress; we then agonize about what we can do to get the upper hand. After running through various scenarios—sometimes over the space of several hours—in search of a means of gaining control over our situation, we give up and go to bed, oftentimes only to remain awake for several additional hours of fretting. On such occasions, we *ruminate*, rather than think, about our problems.

According to *The New Shorter Oxford English Dictionary*, *ruminate* is a transitive verb with a double meaning: either "to chew or turn over in the mouth again" or "turn over in the mind; meditate deeply on." Thus, the word artfully unites the activity of a cow ruminating on a cud and a human ruminating on something purely mental, repetitively turning an idea over and over in the mind.

But self-awareness and self-rumination differ in several important respects. The self-aware person tends to endorse on personality tests such statements as "I'm very self-inquisitive by nature" or "My ideas and feelings about things fascinate me." As a result of such attitudes, highly self-aware people, according to numerous studies, know themselves better than their less self-aware counterparts. They're also more skilled at monitoring and modifying their behavior to complement other people's responses. In addition, self-aware people tend to feel emotions more intensely, act consistently with their expressed beliefs, conform less to social pressure, and express greater empathy toward others. None of this is true of the ruminator, however.

Although the ruminator may also be highly self-aware, his or her emphasis is on the negative: discovering new personality deficiencies and fantasizing about threats and perceived failings. A self-ruminator tends to endorse statements such as

"I'm often focusing my attention on aspects of myself I wish I'd stop thinking about" or "I always seem to be rehashing in my mind recent things I've said or done."

The distinction between a self-aware person and a self-ruminator is important when it comes to anxiety. Even under uncertain and challenging circumstances, the self-aware person is generally less anxious. As a result of increased knowledge about herself, she feels more in control of her future. The self-ruminator, in contrast, has great difficulty accepting that he cannot predict and control the future. What's more, the future always seems threatening. In response, the self-ruminator attempts to manage his anxiety by an even more intense focus on his thoughts—the basis for obsession and compulsion. Think of the obsessive as a die-hard ruminator who aims at gaining an unrealistic and ultimately unattainable degree of control over her life. And, as mentioned previously, *control* is a vital requirement for the obsessive to survive psychologically.

In commenting on the "obsessive style," Salzman wrote: "The fear of loss of control is commonly symbolized in the physical sense as losing control by resorting to the extremes of shouting and screaming, or by engaging in other undisciplined activities. Phobias often develop around these situations, in which the person fears he will lose control of himself in giddiness, fainting, swaying feelings, collapsing, or possibly dying."

In response to these fears, the affected person tends to avoid places or occasions where the dreaded physical symptoms might occur. (Remember Lisa's avoidance of her school and apartment.) In extreme instances, a person afflicted with a phobia may find the imagined loss of control so upsetting that he or she may become a recluse.

The problem of control is further complicated by obsessives' tendency to deal in extremes. Unless they feel in total control, they tend to experience a total *lack* of control. Further, they experience the possibility of loss of control as painfully humiliating, frightening, and dangerous.

While reading Salzman's theories, I thought of just how far we have progressed in our understanding of anxiety. Initially, treatment was directed at providing patients with an understanding of why they act the way they do. Psychiatrists once believed that insight into the condition should bring about a change in the patient's thinking and behavior. But over the years, psychotherapies haven't proven very successful in improving obsessive or compulsive disorders.

After a decade or more of psychotherapy, the OCD patient often remains incapable of restraining himself or herself from checking whether the front door is *really* locked or the oven *really* turned off. Frustrated that their carefully crafted therapies proved ineffective, psychiatrists often evoked such shadowy concepts as resistance—conscious or unconscious efforts by the obsessive-compulsive patient to resist change. But today, concepts like resistance are no longer persuasive.

Recently, the emphasis has shifted toward trying to pin down the anatomy of obsessions and compulsions. During brain imaging, certain areas "light up" whenever the person engages in obsessive or compulsive activity. The hyperactivity occurs in part of the frontal lobes (the orbital frontal cortex), in the anterior cingulate cortex, and (less consistently) in the caudate nucleus, a structure located deep beneath the cortex. After successful treatment, these areas no longer stand out during brain imaging. Such observations have spawned a preliminary theory about how obsessions and compulsions are processed within the brain.

Thoughts and actions are mediated by a two-way circuit that includes the frontal lobes, several underlying structures that comprise what's referred to as the striatal pathway (the name refers to the striated appearance of two of its principal components, the caudate and the putamen) and the thalamus, a collection and distribution center in the brain for all sensations except taste and smell. From the thalamus, impulses return to the original starting point within the frontal lobes, thus completing the cortico-striatal-thalamo-cortical (CSTC, for short) circuit.

One possible mechanism for the intrusive nature of obsessive thoughts might involve a failure in filtering information at the level of the thalamus. Thus, information that is normally processed outside of awareness may, as a result of some disturbance within the striatal components of the network, gain access to consciousness. For instance, as you're reading these words, you're not consciously aware of creases in your trousers or dress. But now that I've mentioned such a possibility, you may find yourself checking for creases and then immediately return to your reading. But if your brain were afflicted with a striatal processing defect, you might not be able to resume your reading because you couldn't get out of your mind the thought of those creases. In response to this obsessive thought, you would be unable to resist the compulsion to check and recheck your clothes. Eventually, you might decide that pressing your clothes is the only way of ridding yourself of this obsession.

Thus, intrusive symptoms (your preoccupation with minor creases in your clothes) bear a direct correspondence to the resulting repetitive behaviors (your compulsive ironing and subsequent reironing of your clothes) that are undertaken to provide some temporary relief from the discomfort resulting from the symptoms. Some researchers even suggest that certain symptom clusters can be localized to a specific component of the CSTC circuit. An obsession with cleanliness, for instance, might be linked with the neural networks concerned with cleaning procedures. The researchers are hoping to eventually establish correlations linking specific obsessions (cleanliness), the brain areas associated with the corresponding compulsions (hand washing), and an effective drug for disrupting the obsessive-compulsive cycle.

But whatever the brain mechanisms responsible for OCD, the person afflicted with the disorder shouldn't consider the situation hopeless. It's possible to retain some control over the symptoms even when in the throes of a particularly agonizing attack. I learned this from the fascinating experiments of Lewis Baxter.

Baxter placed several people with contamination obsessions in a PET scanner and told them that in a few moments they would be handed either a soiled or a clean towel. In reality, all of the towels were clean, but the subjects had no way of knowing that because they could not see the towels while lying inside the scanner. When handed the "clean" towel, brain activation was normal. But with the "soiled" towel, the PET scans showed activation of the OCD circuit involving frontal, striatal, and thalamic structures.

Consider the implications of Baxter's results: Just as one's thoughts can bring on symptoms, an alteration in those thoughts can also dispel those symptoms via an alteration in brain activity. In short, OCD sufferers can obtain relief simply by changing their thoughts about their experiences. Such thought transformation, incidentally, forms the basis for what's known as desensitization treatments.

Take, for example, a person suffering from anxiety associated with an airplane phobia. Several airlines now offer programs for anxious fliers, which involve a gradual exposure to airline travel. The participants in the program begin by sitting on the airplane while listening to short lectures on airplane safety and how favorably airline travel compares to other forms of travel, including their earlier much riskier car trip to the airport. Over several weeks, the participants in the program progress to an actual flight.

Desensitization programs aimed at combating other anxieties work similarly. The person's thoughts and attitudes toward the dreaded object or experience are slowly and gradually changed. The snake is thought of as a potential pet; the doorknob as nothing more than a means of going from one room to another; the bridge as a marvel of architectural engineering that has remained in place for decades. If the treatment is successful, the snake can eventually be handled, the doorknob touched without the use of gloves, the bridge driven across rather than avoided by means of a lengthy detour.

Changing one's thinking is a far different and more effective approach than simply fighting against the anxiety created by the obsession or compulsion. It also works better than someone haranguing OCD sufferers about the unreasonableness of what they're thinking or how they're acting. Intellectually, they already know that they are thinking and acting unreasonably. But such "insight" alone doesn't help. Ask those fighting against an obsession or compulsion, and they will tell you that they fully realize that their avoidances, repetitions, and rituals aren't really necessary. Nevertheless, they feel compelled to perform them in order to restore their peace of mind, for however brief an interval.

What's needed is a modification in the attitude the person takes toward his or her thoughts and reactions. In Baxter's experiments, simply thinking of the towel as clean instead of soiled brought about a change for the better in the brain's patterns. Similarly, the brain's patterns associated with something like a germ phobia can be normalized by thinking of the skin as the normal home of thousands, perhaps millions, of harmless, sometimes even beneficial, viruses and bacteria, rather than thinking of it as a repository of "dirty," "disgusting," or "infective" organisms.

Recently, neuroscientists have focused on learning what takes place in the brain during such thought transformations. In the process, they've come up with a new concept of what it means to forget.

As an example, think of something unpleasant or painful that happened to you as a child that resulted in a temporary conditioned fear. Perhaps you fell off a bike and sustained some minor scratches and bruises. As a result of the fall, you avoided your bike over the next several days because it reminded you of the pain associated with the accident. Eventually, however, you cautiously got back on the bike and, after a few uneventful rides, your fear waned. Such a gradual reduction of fear after repeated uneventful exposures to a previously fearful incident is called *extinction*.

Extinction doesn't involve erasure from your brain of your original fearful memory (the fall); instead, the development of new memories (the succeeding successful "fun" rides) eventually suppress that fearful memory. Thus, if someone asked you several weeks later about your fall from the bike, you would still remember it, but your memory wouldn't be as powerful in shaping your behavior as it had been immediately after the accident. Your new memory of accident-free bicycle riding had overpowered the original fearful memory.

Studies of extinction in humans have highlighted the importance of the prefrontal cortex and the amygdala. "The prefrontal cortex seems to be particularly involved in extinction learning," according to Elizabeth Phelps, a neuroscientist at New York University. Phelps discovered the importance of the prefrontal cortex by concentrating on what happens when people engage in reappraisal: actively shifting their focus and appraisal for an emotional event. Two common phrases capture the essence of the process: "The glass is not half empty but half full" and "Two prisoners looked out through prison bars; one saw mud while the other saw stars." Both phrases employ reappraisal to emphasize the positive rather than the negative aspects of a situation.

To explore the effects of reappraisal on brain activity, Phelps carried out a conditioning experiment on normal volunteers. A brief electric shock to the wrist immediately followed the appearance on a computer screen of a yellow square. Soon the volunteers started tensing up in anticipation of the shock whenever the yellow square appeared. Phelps then instructed them to think of soothing images drawn from nature at the instant they saw the yellow square. In other words, the volunteers were encouraged to reappraise the situation: focus on a soothing image, rather than the fact that the square preceded a shock.

Two exciting findings emerged from Phelps's study. First, the reappraisal strategy reduced the physiological arousal (heart rate and breathing increases) formerly associated with the yellow square—essentially, reappraisal of the situation diminished the fear response. Second, the fMRI results showed an enhancement of the activity of the prefrontal cortex. Another study along slightly different lines carried out by Gregory Quirk of the Ponce School of Medicine, in Ponce, Puerto Rico, found that fear reduction involves prefrontal activation coupled with a reduction of activity in the amygdala.

The bottom line on all this: Anxiety can be mastered by reappraisal. And accompanying this reappraisal, a reconfiguration occurs in the brain's patterning: a shift from amygdalar to prefrontal activation. And this shift enables the anxious person to manage obsessions, compulsions, and anxiety in general in more effective ways.

## A SUMMING UP

Despite their differences, each of the anxiety disorders that we've explored (social anxiety, panic, PTSD, OCD, and agoraphobia and other phobias) represent failed attempts by the sufferers to confine raw anxiety. Each comes at a price. The most costly employ delusions and hallucinations as a means of evading the direct experience of anxiety. Wendell, during the period of his psychosis, is a good example of this escape-from-anxiety via an escape-from-reality approach. The overpowering anxiety, created by the memory of his train killing the suicidal woman Joan, eventually so overwhelmed Wendell that he could only obtain relief via the psychotic conclusion that Joan wasn't really dead. But even this failed to restore his peace of mind, since Joan reappeared in his visions and threatened to kill or harm him.

In the other anxiety disorders, insight is retained to a certain extent: Lisa didn't actually believe that she hadn't turned in her test paper, or that she hadn't paid her rent on time and would therefore be evicted from her apartment. Rather, her anxiety resulted from her inability to *stop thinking* those distressing thoughts. Like Lisa, sufferers from OCD, social anxiety, and GAD recognize that their anxious thoughts and feelings aren't based on realistic concerns. As a result of the anxiety created by the overactivity of their brain's emotional circuitry, they see the world through a pair of colored lenses that render everything in forbidding monochrome.

# 8

## "Life Is a Daring Adventure or It Is Nothing"
### *Defying Fear and Anxiety*

*E*VEN THE MOST anxious person isn't necessarily anxious all the time. Rather, the majority of people become anxious only in response to special circumstances, especially in the face of uncertainty.

In the 1960s, psychologists R. B. Cattell and I. H. Scheier pioneered a new approach to the study of anxiety aimed at distinguishing between a tendency toward anxiety and the occurrence of actual anxiety episodes. Starting with self-reports from anxiety sufferers, Cattell and Scheier correlated the descriptions with physical measures such as pulse rates and blood pressure. Two related but different categories emerged: an unpleasant, painful emotional condition that fluctuates over time and varies in intensity, which they dubbed state anxiety (shortened to S-anxiety), along with a more general propensity toward periodically experiencing anxiety, which they referred to as trait anxiety (shortened to T-anxiety). Think of state anxiety as the actual experience of anxiety, whereas trait anxiety is the tendency to develop anxiety episodes. Not surprisingly, Cattell and Scheier found that people with high levels of T-anxiety frequently slip over the edge and experience intense elevations in S-anxiety.

During the intervening forty years, the trait-state distinction formulated by Cattell and Scheier has provided the inspiration for the work of a man informally referred to by his colleagues as Mr. Anxiety. At seventy-seven years of age, Charles Spielberger is presently Distinguished University Research Professor of Psychology at the University of South Florida. The author of more books and papers on anxiety than anyone alive, his works include *Understanding Stress and Anxiety* and the four-volume *Cross-Cultural Anxiety*. But he is most renowned among his colleagues for the Spielberger State-Trait Anxiety Inventory. In order to learn more about the state-trait distinction, I talked with Spielberger, a scholarly appearing man who speaks in a measured cadence and a soft accent.

"Anxiety is built into us because it's adaptive and provides us with the motivation to escape and survive," he said. "It focuses our attention on problems so that we can solve them. But anxiety isn't the same for any two people. When I speak of an anxiety *state*, I'm referring to the anxiety that is experienced at a particular time. It corresponds to our common sense concepts of anxiety: feelings of tension, apprehension, nervousness, and worry. Those feelings vary in intensity from time to time and from person to person. *Trait* anxiety is a disposition to envision the world as a dangerous place. It's a free-floating tendency to sense threat in objectively nonthreatening situations."

As Spielberger spoke, I thought of Lisa. At the time of her second visit to my office, Lisa was in an anxiety *state*. But this improved over the next two weeks after the prescription of a drug, Paxil, that modified her brain's neurotransmitters, particularly serotonin. Here is Lisa's description of how her anxiety shifted from state to trait and, finally, to normal psychological functioning.

"All those horrible thoughts are now gone. Looking back, it's like I was living in an alternate reality. No matter how illogical and crazy things seemed at the time, those were my actual beliefs. Do you remember when I told you about the exam paper and my fears that I hadn't handed it in? Even as I was telling you that I realized how crazy it all sounded. Yet in my state of mind at the time, I really believed that the paper had been mislaid during the few seconds that elapsed between my handing the paper to the monitor and returning to my seat.

"What's more, at that time I didn't have any confidence that things were going to change or that anything I could do would help. In fact, when you gave me the prescription, I didn't believe the drug would work. I just *couldn't imagine* feeling any differently from how I felt that day. As a result, when you suggested that I would feel better by the time of my next visit two weeks later, I just didn't believe you. *He can't have any idea how I feel,* I remember thinking at the time. Yet a week later I had the wonderful surprise of feeling so much better. It was like I was entering a new world. I can't tell you how great it feels to be back in charge of my life once again after feeling out of control for so long. Yet nothing objective has changed to explain why I feel better. For instance, the results from the exam that I was so worried about won't be returned for another week. But I'm not bothered about that. I was well prepared for the test, and I'm confident I did fine. But even if I didn't do as well as I'd like, I can now handle the stress of getting a lower grade just as long as I passed. Most wonderful of all, I no longer have that feeling, that dread, that something horrible is about to happen and I'm going to go under."

Notice Lisa's description of the change in her inner experience. That change is so striking that she can no longer emotionally experience how she felt at the time of her first visit when she was in an anxiety *state*. Based on the fundamental difference between her present and past states, it's as if the Lisa of today is describing an earlier and different Lisa she cannot empathically identify with. Novelist Padraic Colum described this dilemma in a passage from his novel *The Flying Swans:* "No one knows the state one is going into until one is in it. No one knows. No matter how one fancies it, it is always different."

Will Lisa be at risk for future episodes of intense anxiety? I believe that she will because even though her anxiety state is now in remission, she retains an anxiety trait—that is, a tendency to develop state symptoms under specific pressures. And since Lisa has experienced both her anxiety episodes just prior to midterms, it's likely that a similar flare-up could occur next year at the same time.

Spielberger believes that anxiety is especially likely to occur in situations like Lisa's, when the individual feels under scrutiny or evaluation. Perceived criticism or negative evaluations—or even the prospect of such developments—can lead to high levels of both state and trait anxiety. "People with high trait anxiety have many more anxious experiences stored in their memory; consequently, any situation in which they feel they are being criticized or found wanting in any way evokes memories of similar reactions in the past," he said. "This enhanced memory for past anxiety episodes increases the amount and intensity of the present anxiety experience."

BUT NOT EVERYBODY REACTS to uncertain, ambiguous, or potentially threatening situations by experiencing anxiety, developing panic attacks, or hiding under the bedcovers. Some people even enjoy the anxious anticipation of potential catastrophes and go out of their way to court them.

Psychologist Frank Farley specializes in the study of risk takers. He believes we all exist somewhere along a continuum ranging from risk taking to risk avoidance.

At one end of the continuum are the *small t* (with the *t* standing for "trait") individuals who are adverse to risk of any sort. Typically, the small t individual restricts his or her excitement to the occasional wild amusement park ride or horror movie.

At the other extreme of the continuum are the risk takers—the *Big Ts*, as Farley refers to them—the thrill-seeking, arousal-driven "adrenaline junkies" who thrive on the anxiety associated with uncertainty, novelty, variety, and intensity. The Big T willingly, even enthusiastically, puts himself or herself in physically dangerous situations like hang gliding, bungee jumping, extreme skiing, or big-wave surfing. "Big Ts respond to fear by taking action and thrive in a world of unceasing novelty, unpredictability, and absence of structure," said Farley.

Typical of a Big T is extreme skier Kristin Ulmer: "It takes a tremendous amount of stimulation to get me excited. I love excitement, whether it's meeting a new person or jumping off a cliff, I love it. It makes me feel alive."

A *National Geographic* map titled "The Quest for Everest" shows not only the relevant geography but also an illustrated history of mountain climbers ("Stout Hearts and Hubris on the Highest Mountain"). While there have been many successes, many others never completed the climb. Since the May 29, 1953, initial ascent by Edmund Hillary and Tenzing Norgay, more than 170 have died, with some 120 corpses remaining on the mountain. Why do people deliberately expose themselves to such hazards? Certainly the reasons differ from one person to another.

- For Reinhold Messner, it was the feat of reaching the pinnacle alone in 1980 with no partners, no porters, no radio, no rope, and no supplemental oxygen.
- For Stacy Allison, the goal in 1988 was to become the first American woman to reach the summit.
- For Peter Hillary in 1990, it was to match the achievement of his father.

Contemporary mountaineer Kim Clark, a thirty-four-year-old nurse from Denver, Colorado, is inspired mainly by internal rather than external goals. She takes satisfaction in "trying to be the best that I can be . . . and who knows what that is? I just keep trying and it feels good!"

Notice that these Big T contemporary mountain climbers don't dwell on the risks of death or injury. They also respond differently from most of us to physiological arousal. For them, the flood of adrenaline accompanying a potentially dangerous venture elicits pleasure rather than anxiety. The Big T is driven by the desire for internal mastery.

For other Big Ts, the risks may be mental rather than physical, as with the high-stakes venture capitalist who thrives under the uncertainty of entrepreneurship, or the politician fighting to bring about his or her version of government reform, or the social activist willing to champion unpopular causes.

"The risk-taking can also be intellectual," Farley claimed. "Albert Einstein was thrilled by his mental life. The thrill of uncertainty kept him going. The same dynamic was at work with Winston Churchill or Pablo Picasso. These were not people who choked while standing on the edge."

But how does one separate your garden-variety Big T from someone who is simply reckless or even self-destructive? A small t individual, of course, may refuse to recognize such a distinction. To the small t, the risks willingly undertaken by his or her Big T counterpart aren't acceptable under any circumstance. "Why would anyone willingly expose themselves to unnecessary risks, especially under situations where the reward is only an internal one?" the small t asks. Only recently has that question been addressed.

"Much of the research on the fear response in humans has focused on its dark side—that is, the multiple ways it can get stuck, out of whack, or bollixed up: irrational anxiety, panic attacks, obsessive-compulsive disorders, post-traumatic stress. Viewing fear as only pathology it's sometimes easy to forget how adaptive and essential a part of life most fear truly is," said author Jim Thornton, who wrote "The Joy of Fear," which was featured in *National Geographic Adventure* in its June–July 2002 issue.

Thornton and Farley are no doubt correct that many of us are overly fearful and anxious. But, in fairness, I think some heed should be given to the advice of the nineteenth-century statesman Edmund Burke: "Early and provident fear is the mother of safety." Imagination is often the distinguishing feature separating appropriate from excessive and inhibiting anxiety. A person's capacity for imagining various danger scenarios influences his or her attitude toward danger. An uneasy balance is called for between fearlessness and lack of imagination. Too powerful an imagination concerning risks leads to paralyzing timidity; too weak an imagination, and one fails to anticipate possible harmful consequences.

For example, a neurologist friend recently decided to do some skydiving during a trip to Hawaii. When I spoke to him several hours after the experience, he appeared shaken.

"Rather than taking skydiving lessons, which would have consumed several weeks, I elected to have myself tethered to an experienced diver," he said. "We jumped from the plane at sixteen thousand feet and remained in free fall for fifty-five agonizing seconds before my partner opened the parachute. I never imagined the experience would be anything like that total loss of control and total vulnerability."

"Would you do it again?" I asked him.

"No, I wouldn't. I guess I just never imagined anything like it. And I certainly hope I won't have any dreams re-creating how I felt during that fall."

As a doctor, I've encountered more tragic examples resulting from failures of the imagination: failures to take into account both the good and bad things that can happen. Included here are paraplegic and brain-injured young adults incapacitated by accidents incurred while engaging in one or another extreme sport or risky endeavor. Talking with them, I've had the unsettling impression that prior to their accident they simply had never imagined that their lives could take such a tragic turn. Their fear (which they suppressed or ignored) had become somehow disconnected from the thrilling though dangerous activity they were engaging in: "It couldn't happen to me" was their unspoken motto. But one day it did. When discussing with them the circumstances leading up to their injuries, I've found it hard to resist the conclusion that their injuries resulted from an inability to mentally envision the tragedy that eventually befell them.

Frank Farley didn't agree with me during our conversation that Big Ts are less imaginative than their small t counterparts: "The Big T isn't just a reckless daredevil who denies the possibility of death or serious injury, but a person with enormous self-confidence who has learned to overcome challenges by taking one small step at a time. In addition, the Big T is a creative flexible thinker who loves choice, variety, change, and challenge."

Farley makes two recommendations for bringing about an increase in one's fear tolerance. The first is summed up in a favorite quote of his from Ralph Waldo Emerson: "Always confront the things you are afraid of." Another quote he especially likes, because it makes a similar point, comes from Helen Keller: "Life is a daring adventure or it is nothing."

The daring adventure is overcoming fear and anxiety so that life can be experienced. While Keller could have lived a life imprisoned in her body, she chose freedom and reached out to the world. Ninety-nine percent of the things people fear turn out to be paper tigers once they're confronted, according to Farley. What's needed is the determination to take specific steps to overcome those fears.

As an example, Farley cited a former professor afflicted with a paralyzing fear of spiders. The fear evaporated when the professor cultivated an interest in spiders and learned more about them: "That professor changed his fear of spiders into an intellectual passion. Instead of living in dread of spiders, he sought to know everything there is to know about them. And after he made that inner change in attitude, spiders went from objects to be feared to interesting creatures that he could keep learning more and more about."

Labeling has a lot to do with our anxieties and fear, according to Farley. "It's our *interpretations* of events rather than the events themselves that cause the problems. Like that professor, you have to force yourself to take specific steps to overcome your anxieties and fears."

As a first step in changing interpretations, the anxious person must be willing to engage in some self-analysis. As a child, extreme skier Ulmer experienced a wide range of fears that she has spent a lifetime overcoming by analyzing one fear at a time: "Ask yourself, 'Are my fears irrational? What really could happen here?' If you're afraid while you're doing something risky, you're going to get hurt. But if you face your fears, whether it's something as simple as a fear of heights or as complicated as how to rock climb, the rewards are tremendous."

While Ulmer is undoubtedly correct about the importance of facing one's fears, I suspect that as a result of her skiing experiences, she is no longer capable of putting herself "in the skin" of a small t. After years of encountering risks and managing her fears, Ulmer, like most Big Ts, gets euphoric in response to experiences that would reduce the average person into a whimpering puddle of mush.

Most of us dwell somewhere toward the middle of Farley's continuum: We want sufficient novelty and excitement in our lives to fend off boredom, but we're also attracted to certainty, safety, and routine. And I suspect there's a limit—genetically determined, mostly—to how fearless a small t can ever become.

Frank Farley confirmed my doubt that a small t can ever transmogrify into a Big T, no matter how determined he or she is to make the change: "It only happens in fiction. Clark Kent goes into a phone booth as a nerdy, anxious, low-risk taker and a moment later emerges as Superman, who isn't afraid of anything. In real life, however, such drastic personality changes just don't happen. A small t can advance himself in small steps along the type T continuum, but he isn't going to be able to become a Big T. In fact, even trying to make such a radical change may be unbearably stressful for the small t."

But even incremental increases in a person's anxiety tolerance can make a big difference. "Anxiety drives people to do terrible things," Farley continued. "The anxious person is prone to irrational fears about other people and their motives. This can lead to false judgments and irrational actions."

Farley's comment about false judgments and irrational actions resulting from fears and anxiety occurred to me recently while listening to an NPR interview of an expert on North Korea. The expert made the point that President Bush's comments on several occasions about preemptive war ("America will act against emerging threats before they are fully formed") created "anxiety" among the North Koreans that the United States might launch a nuclear attack, thus justifying in the minds of the North Koreans the development of their own nuclear capability. Such an attitude seems puzzling to most Americans, who strongly believe that our country would never launch a preemptive nuclear strike. But, obviously, leaders in other nations don't know us as well as we know ourselves. As a result, they aren't certain of our intentions—and that uncertainty breeds anxiety.

"It is important to ask ourselves, as citizens, whether a world power can provide global leadership on the basis of fear and anxiety," wrote Zbigniew Brzezinski, a former national security adviser, in the *Washington Post*. What can we anticipate if we pursue a foreign policy based on creating and sustaining a state of anxiety among other nations throughout the world? Do nations, like individuals, take unwise and ill-considered actions when the collective anxiety in the population reaches a certain threshold? For instance, if the leaders in Pakistan become sufficiently anxious at the prospect of neighboring India launching a nuclear attack, could that anxiety propel them toward a preemptive strike?

Although we don't usually place anxiety among the first half dozen or so determinants of whether a nation goes to war, the universal experience of anxiety is a prime determinant of international relations. Insecurity, at both the personal and the collective level, forms the basis for this anxiety, according to Guido den Dekker of the Department of International Law, University of Amsterdam. "International security ultimately depends on the security of individuals," den Dekker wrote in a paper entitled "From Human Insecurity to International Armed Conflict."

Geopolitically, anxiety increases in direct proportion to the sense of insecurity felt by individuals within a nation.

"One of the root causes of war lies in the process which leads from a situation of individual human insecurity in different areas of life to instability of society and from there to international reactions including the use of armed conflict. . . . Fears and perceptions of threat by other states may give rise to hostile international reactions affecting the already unstable society," wrote den Dekker.

Currently, one of the greatest sources of collective anxiety and the resultant destabilization is the threat of terrorism. "CIA Says al-Qaeda Ready to Use Nukes" is a headline taken from a recent *Washington Times* article. "Terrorists are set to use chemical, biological and nuclear weapons . . ." is the equally anxiety-arousing sublead for the article. While such information is helpful and necessary when placed in context, the wording arouses anxiety. Nor does the article, based on a CIA report entitled "Terrorist CBRN: Materials and Effects" (CBRN stands for chemical, biological, radiological, and nuclear), provide any specific steps that the average reader can take to prepare for—much less prevent—such developments. Further, the article and the report on which it's based warns that "such attacks cannot be ruled out" even though "no information proves the group now is planning an attack in the United States with weapons of mass destruction." In short, the article raises the reader's anxiety about something dire that may or may not occur at some indefinite time in the future and, further, something about which he or she can do little.

"This may be the only time in our history when we are not only warned that we should be afraid, but told exactly how afraid we should be (red, orange, or yellow alert), and yet, regardless of how afraid we should be, we are given no advice about what to do, except perhaps to be wary of strangers and stock up on duct tape and bottled water. What is the effect of this?" asked the composers of a thematic statement introducing a symposium called "Fear: Its Political Uses and Abuses." Under such circumstances, anxiety becomes an inevitable response both individually and in society at large. And as Al Gore mentioned in his speech to the Fear symposium, "When fear is activated, it is very difficult to turn off." Although Gore and other symposium speakers spoke of fear, the actual day-to-day experience they describe is anxiety: a widespread, nameless, uncomfortable feeling that something is dreadfully wrong or may go wrong at any moment.

Communal anxiety may be expressed in different forms, according to den Dekker. The three most common forms are civil unrest, ethnic tensions, and social disintegration and fragmentation. Judging from the daily papers and the evening nightly news, we're already deeply immersed in the communal anxiety described by den Dekker. And although this book isn't about political science, a discipline in which I claim no professional expertise, I think I'm on solid ground when I say that life in the twenty-first century provides and will continue to provide innumerable examples of the three forms of communal anxiety described by den Dekker.

# 9

## "Something I Had Never Seen Before"

*Antianxiety Drugs*

C URRENTLY, A LOT OF PEOPLE respond to stress by reaching for anxiety-relieving medications, which generate millions of dollars per year in revenue. In 2001, Americans spent $715 million on tranquilizers such as Valium and Xanax. And stressful events typically lead to an escalation in tranquilizer use. For instance, in the two weeks following the World Trade Center attack, prescriptions for Xanax increased nationwide by 9 percent (and by 22 percent in New York) before the usage patterns gradually returned to pre-terrorist-attack levels. How did the use of tranquilizers to quell anxiety become so deeply ingrained in our culture? To answer that question, it's helpful to look at the history of anxiety-relieving substances.

Prior to the 1960s, no agents existed that were capable of selectively diminishing feelings of anxiety. Of course, opiates, alcohol, and barbiturates could temporarily reduce anxiety, but the usefulness of these compounds was limited by serious problems that developed during long-term use. All three agents produce dependence, vary widely in their effects from one day to another (ask any alcoholic), and can prove dangerous when taken in excess. Not surprisingly, researchers were on the lookout for alternatives. Unfortunately, they hadn't a clue about which chemicals to test in the search for safer, more effective antianxiety agents. As often happens under such circumstances, the combination of chance and a prepared mind led to the pivotal breakthrough.

The first tranquilizer was an unintended by-product of research in the late 1940s aimed at developing an antibiotic capable of killing penicillin-resistant bacteria. After the synthesis of the first such antibiotic, the lead researcher, Frank M. Berger, tested it for safety by injecting the drug into mice. After the injection, he observed

> something I had never seen before . . . The compound had a quieting effect on the demeanor of the animals. They lost their righting reflex so that they were unable to turn over when put on their back. Their muscles were limp and completely relaxed. Yet the animals appeared conscious. Their eyes were open and they appeared to follow what was happening around them. After the paralysis, which lasted for a period ranging from a few minutes to a few hours—depending on the dose of the drug—there was spontaneous and complete recovery to the state the animals were in prior to the administration of the drug. This effect was described as "tranquilization."

Berger and his team decided to evaluate other chemicals based on their ability to produce loss of the righting reflex in mice. Eventually, they came up with a chemical they labeled Miltown (adapted from the New Jersey town of Milltown, the home of Wallace Laboratories, the manufacturer) and subsequently given the generic name meprobamate. The most impressive feature of the drug was its taming effect: formerly vicious monkeys given Miltown could be easily handled.

Released for prescription use in 1955, Miltown sales catapulted from $7,500 in May to more than $500,000 in December. Soon sales exceeded $100 million a year, thanks in part to magazine articles about the drug with catchy titles such as "Happy Pills," "Peace of Mind Drugs," and "Happiness by Prescription." By 1957, Miltown had become so well known and its use so prevalent that comedian Milton Berle began introducing himself to his TV audience with "Hello, I'm Miltown Berle." As a result of the drug's burgeoning popularity, the stigma that usually attends the use of a "psychiatric drug" vanished. After all, how could there be a stigma attached to a drug "everyone" was talking about, taking, or thinking about taking? For the first time in history, the mass treatment of anxiety in the general community seemed possible.

The huge success of Miltown spurred other pharmaceutical companies to come up with an equally successful drug. But the competing drug had to conform to specific criteria: It had to be at least equally effective, produce fewer side effects, and differ sufficiently from Miltown in its chemical composition that it could be patented as a separate drug. Once again, luck played

a major role in the development of the tranquilizer that suc-
ceeded Miltown.

While searching for a drug with sedative properties, chemist
Leo Sternbach modified a chemical originally developed as a
dye. He found the chemical to be ten times more effective
than Miltown as a muscle relaxant. Further, the chemical
worked wonders in what experimental psychologists refer to as
an "induced conflict situation."

The conflict involved hungry rats trained to press a lever for
food. (As you've no doubt observed by this point in the book,
the lowly rat has certainly done more than its share of the
"heavy lifting" when it comes to research on anxiety and
antianxiety drugs.) After the rats learned to push the lever at
a steady rate, the experimenter turned on a light and kept it on
for a few minutes before turning it off. When the rats pressed
the lever in the presence of the light, they received a mild
shock as an accompaniment to their food. After a few trials,
the conflict situation was established: The rats hesitated to
press the lever for food when the light was on. They also
showed signs of anxiety (excessive grooming and defecation).

At this point, Sternbach gave the rats the new compound.
Within minutes, the animals resumed pressing the lever even
in the presence of the light and the mild shock. And since
morphine and other narcotics didn't produce the same effect,
Sternbach sensibly concluded that the compound didn't work
by simply reducing the rat's pain sensitivity. Rather, the chemical reduced the animal's "anxiety" about the shock.

Although not everyone equated signs of restlessness in a rat
with anxiety in a human, the success of this new compound
(named Librium; the chemical name is chlordiazepoxide) in

alleviating the rat's dreaded anticipation of a repeat electric shock suggested a hypothesis for the development of future tranquilizers: Any compound that reduced signs of anxiety in an animal (an anxiolytic) would also reduce human anxiety.

Librium's success was furthered by Miltown's rapid fall from grace. It turned out that the world's first and most popular tranquilizer produced both psychological and physical dependence. As a result, Miltown yielded its place in 1960 to Librium, which within a few years achieved the status of the world's most prescribed tranquilizer.

Despite Librium's effectiveness in reducing anxiety, scientists hadn't a clue about how the drug worked within the brain. The pivotal insight into its mechanism of action came in 1975, when researchers discovered that Librium and other soon-to-be-developed drugs of the so-called benzodiazepine family augment the actions of a neurotransmitter called gamma-aminobutyric acid, or GABA for short. As an aid to understanding the action of GABA, here is a three-paragraph summary of how neurons communicate with each other.

Basically, neurons communicate by means of special chemicals, called neurotransmitters. As strange as it may seem, all mental processes result from the release of neurotransmitters from billions of cells in the brain and the reception of these chemicals by billions of other cells. This process of neurotransmitter release and reception takes place at the synapse, the site where nerve cells come into close approximation with each other. The tiny space separating two neurons is called the synaptic cleft.

When the nerve impulse carried by the messenger neuron (referred to in technospeak as the presynaptic neuron) reaches the synapse, it releases a chemical, the neurotransmitter, which then flows across the synaptic cleft and locks on to its specific receptor on the surface of another neuron (the postsynaptic neuron). This binding of the neurotransmitter to its receptor alters events within the interior of the postsynaptic cell, setting in motion a cascade of chemical processes. If these processes lead to activation (excitation), the neuron "fires," leading to a high-velocity electrical impulse that conducts the nerve "message" to the next synapse. The widespread replication of this message within neuronal circuits results in information transfer throughout the brain. If, instead, the processes within the neuron lead to deactivation (inhibition), the neuron fails to fire, and no impulse is transmitted.

The take-home message is that both normal and abnormal thoughts, emotions, and perceptions depend upon the integration of these excitatory and inhibitory signals in the trillions of synapses within the brain. Too much inhibition leads to stupor or coma. Too much excitation results in seizures or other conditions marked by overexcitation, such as anxiety. It all depends on the interplay of the neurotransmitters.

Of all the neurotransmitters (numbering about seventy-five, although no one knows exactly how many exist), GABA plays the main role in the action of drugs aimed at reducing anxiety. The drugs work by activating one of several sites on the GABA receptor. Barbiturates activate their own site on the GABA receptor, as do the benzodiazepines (commonly referred to simply as "benzos"). As you might expect, GABA receptors are especially concentrated on the neurons of the cerebral cortex and the subcortical areas involved in the experience and expression of emotions.

Activation of the GABA receptors by a benzo leads to mild inhibition of the neuronal firing rate. This is similar to but less powerful than the inhibition that results from other GABA-activating drugs such as sedatives, antiepileptics, and hypnotics. Think literally. Since excitement, epilepsy, and sleeplessness result from neuronal overactivity, drugs that decrease (inhibit) neuronal firing can improve each of these conditions. Likewise, since anxiety results from overexcitation within the brain, especially within those circuits mediating emotion, drugs such as benzos that dampen this overexcitation are anxiety relieving or anxiolytic (a combination of *anxiety* and the Greek *lytikos*, "able to loose").

A decade after the introduction of Librium, another benzo arrived on the scene. Valium, a more potent (lower dose), more rapidly effective agent went on to became the leading seller among all prescription drugs. In 1973, its peak year, Valium sales in the United States reached $230 million, or approximately $1 billion in current dollars when adjusted for inflation. And while Valium proved extremely powerful in relieving anxiety, it turned out, like Miltown, to be habit-forming. As described in the Rolling Stones song "Mother's Little Helper," many people were "running for the shelter" of a pill that would enable them to get through their day.

According to one estimate in the 1970s, there were 10 million Valium addicts in the United States. But despite its addiction potential, Valium and the other benzodiazepines remain to this day the most commonly prescribed drugs in the world. Included in this family are a variety of drugs sharing a common chemical structure. They vary principally in the time required for the onset of the antianxiety effect, duration of action, and potency. But despite variations from one drug to another, the benzo-GABA interaction resulting from the use of these drugs illustrates an intriguing symmetry: The decrease in excitation at the molecular level brought about by the tranquilizer is mirrored at the behavioral level by a decrease in anxiety. Placation of the "anxious" neuron induces tranquilization in the anxious person.

AT THE MOMENT, antidepressants, especially the serotonin reuptake inhibitors (SRIs) such as Prozac, Paxil, and Zoloft, have replaced the benzos as the first-line treatment for anxiety. These newer, safer, and less habit-forming drugs work by prolonging the time spent by a neurotransmitter within the synapse. They accomplish this by blocking a normal process whereby the presynaptic neuron sweeps up the neurotransmitter and reprocesses it. For reasons that remain incompletely understood, depression improves in response to this inhibition of the reuptake of serotonin from the synapse (hence the term *serotonin reuptake inhibitors*). Similar antidepressant effects result from inhibitors of norepinephrine and other neurotransmitters.

For instance, selective serotonin reuptake inhibitors (SSRIs) were initially used only for depression. Additional experience with these drugs led to the observation that the drugs worked equally well with anxiety. This observation raises an interesting question that is presently generating a great deal of controversy among psychiatrists: Are anxiety and depression somehow related?

Traditionally, psychiatrists considered anxiety and depression as separate entities. But following the discovery that anxious patients often responded to antidepressants, psychiatrists were forced to explain the unanticipated effectiveness of anti-depressants as antianxiety agents. Although no one so far has come up with a completely satisfying explanation, several speculations are currently popular.

Some psychiatrists are suggesting that anxiety and depression exist along a continuum. At one end of the continuum is endogenous depression, whose sufferers fit the popular stereotype of depression: melancholy, inactive people who get little pleasure in life, maintain few interests, and seem totally lacking in energy or enthusiasm. As the word *endogenous* implies (*endo*, from the Greek *indon*, "within"), this form of depression results from internal rather than external life experiences. Typically, people with endogenous depression also voice many physical complaints, such as insomnia, decreased appetite, and a general lack of energy.

Further along the continuum is exogenous depression, which is caused by external circumstances (from the Greek *exo*, "outside") rather than internal psychological factors: somebody dies, your spouse wants a divorce, your kid is into drugs. Exogenous depression is typically characterized by restlessness, indecisiveness, obsessions, and fretfulness. Finally, at the opposite end of the continuum are the anxiety states, which share many of the traits of exogenous depression.

Despite its usefulness, the continuum model fails to take into account the fact that anxiety and depression frequently occur together. For instance, somewhere between 50 percent and 80 percent of people with GAD have one or more other disorders, usually depression. Such associations suggest that anxiety and depression may share "biological vulnerabilities," perhaps something like an overactive response to stress.

Others speculate that shared genetic features play a less-important role than environmental stresses when it comes to determining whether a person will develop anxiety or depression. For instance, anxiety leads to social isolation, which results in depression because of the consequent lack of interaction with other people; depression, in turn, creates anxiety due to the gloom and foreboding resulting from apprehension that the sad feelings will never end.

Nor is it always easy to distinguish depression from anxiety. People with both conditions often complain of the same things: an inability to relax, a sense of impending doom, and difficulty concentrating or even sitting still for more than a few moments at a time. But despite these similarities, making the correct diagnosis and applying the correct treatment can sometimes be a matter of life or death.

Tranquilizers like Valium typically lessen anxiety but worsen depression. Thus, mistreating depression with a tranquilizer rather than an antidepressant increases the chances that the depressed person may try to kill himself or herself. Indeed, this increased risk of suicide among depressed people mistakenly diagnosed as suffering from anxiety is one of the reasons why psychiatrists select an antidepressant rather than a tranquilizer whenever they're in doubt about the correct diagnosis. Happily for both patient and doctor, this anxiety versus depression dilemma has been eased somewhat since the discovery of the effectiveness of the selective serotonin reuptake inhibitors (SSRIs) in combating both conditions.

Unfortunately, SSRIs aren't immediately effective; it takes about three weeks before patients begin feeling better. Why this delay? It's likely that serotonin isn't the only culprit responsible for anxiety. Therefore, drugs aimed at influencing serotonin may actually work by restoring a normal balance among many other neurotransmitters. And this typically takes weeks. Such a delay also suggests—and there is good evidence to support this—that the drug is influencing the previously mentioned cascade of chemical reactions occurring within neurons.

Think back to the last time you turned the key on the ignition of your car and the car wouldn't start. At that moment, the possible causes for this malfunction ranged from defects in the ignition, to your failure to notice a day earlier that the car was almost out of gas, to a host of mechanical problems that more than likely will require a mechanic to sort out. A similar uncertainty about causation exists in the brain when it comes to neurotransmitters.

Serotonin-altering drugs act by favorably modulating processes in the brain that aren't directly linked with serotonin. Alternatively, increasing the amount of serotonin may be nonspecific: Altering other neurotransmitters might be equally effective in restoring a normal balance. Think of all of the neurotransmitters as functioning like the instruments in a symphony orchestra: A flawed performance by any one instrument can produce an overall deterioration in the performance of the entire orchestra.

Future drug development aimed at the treatment of anxiety is likely to differ dramatically from earlier efforts. When Leo Sternbach synthesized Librium and later Valium, it took less than three years to carry out the clinical phase of drug testing. It now takes six or more years for an antianxiety drug to be tested for safety and efficacy. And instead of working on one drug at a time, drug developers now rely on computers to generate many hundreds of candidate molecules that are then screened for their potential anxiolytic effect on humans.

Finally, advances in molecular biology and the understanding of the human genome will drastically alter the traditional method for developing new drugs. Instead of a "let's try this chemical and see what it can do" approach, drug companies are already identifying and "attacking" specific molecular targets within the brain—a process described by psychopharmacologists as "rational drug design."

Another approach involves administrations of chemical "probes" that are specific for individual neurotransmitters and their receptors. Injection of one of these probes provides a color-coded map of the distribution of the receptors for a specific neurotransmitter. Once an abnormality in the distribution pattern is identified, researchers attempt to come up with a drug that will correct the situation. If they've planned carefully enough, and are lucky enough, correction of the chemical abnormality may result in a successful treatment for an anxiety disorder or other neuropsychiatric condition. As an accompaniment to this research, it's likely that researchers will encounter new and surprising insights.

While chemicals favorably influence anxiety in people afflicted with one of the anxiety disorders mentioned earlier, mild anxiety can often be managed effectively by taking sensible measures in regard to diet and nutrition (easy on caffeine and other stimulants); sleep (increase your chances of falling asleep by relaxing during the last few hours in the evening rather than fretting about your activities for the next day); and exercise (even if you're a confirmed couch potato, you still have to do *some* walking everyday, so why not do more of it?). In addition, there are two reasons why the use of such medications isn't a sensible response to the everyday anxiety that we all experience from time to time.

First, the anxiety usually returns after the medications are discontinued. To this extent, critics of the psychopharmacological approach to mental health have a point. If a person feels free from anxiety only while on a medication—assuming that she isn't suffering from one of the anxiety disorders mentioned earlier—then a change is required either in her external circumstances or in her attitude toward those circumstances. And since any one of us can exert only a limited influence on the cataclysmic changes taking place in the world that are presently contributing to our anxiety (terrorism, economic downturns, events in our personal lives, etc.), one's attitude becomes increasingly important. Recognize that some anxiety is normal under the circumstances under which we are forced to live. In less than five years, we have gone from a confident, powerful, comparatively carefree superpower nation to one that has been forced to experience a newfound sense of vulnerability. Anxiety is also normal in response to personal events: for example, the "jitters" we all experience during the weeks preceding a wedding, or the agonizing uncertainty we feel about whether we will be promoted at work. It's anxiety relieving to accept those things that we cannot change.

Second, anxiety is a lot like pain—it sometimes serves as a warning that something is amiss and needs our attention. Physicians, with good reason, are reluctant to be too aggressive in medicating away a patient's pain prior to establishing the reason for the pain. That's because the patterns and rhythms of the pain provide important clues about what's causing it. Similarly, medicating anxiety away may deprive us of the opportunity to learn its origin.

Notwithstanding the above considerations, tranquilizers and other anxiety-relieving medications can, when used sensibly, effectively relieve the most painful and incapacitating aspects of anxiety.

Perhaps Frank Berger best summed up the benefits of antianxiety agents when properly used: "It would be wrong and naïve to expect drugs to endow the mind with insights, philosophic wisdom, or creative power. These things cannot be provided by pills or injections. Drugs can, however, eliminate obstructions and blockages that impede the proper use of the brain. Tranquilizers, by attenuating the disruptive influences of anxiety on the mind, open the way to a better and more coordinated use of the existing gifts. By doing this, they are adding to happiness, human achievement, and the dignity of man."

## A SUGGESTION FOR TRANSFORMING YOUR ANXIETY

*If Necessary, Supplement Your Own Efforts at Managing Your Anxiety by Means of Medications*

If your anxiety occurs in the absence of any specific themes, this suggests the presence of a medical or psychiatric disorder. Make an appointment for a general physical checkup. If everything turns out fine on the exam, your anxiety is probably arising from one of the anxiety disorders described earlier in this book. Consult with a psychiatrist or other specialist physician skilled in the diagnosis and treatment of anxiety disorders. He or she may suggest a medication.

Anxiety-relieving medications reduce anxiety from disabling to tolerable levels. When taken at the correct dosage, frequency, and duration of treatment, these drugs aren't habit-forming. Since they take effect immediately and remain effective for only a few hours, tranquilizers usually must be administered several times a day. Some people with overwhelming anxiety require daily dosing ranging from two to four pills, depending on the drug chosen and the intensity and frequency of the anxiety. People with less-frequent, less-intense anxiety don't require medication every day, but can achieve satisfactory control by taking medication only as needed. But as-needed dosing doesn't work with drugs such as the SSRIs, whose effectiveness depends on achieving an alteration of the neurotransmitter balances throughout the brain. In order for these drugs to be effective, they must be taken daily over a long-enough period (usually three or four weeks) before the drug reaches what's called a "therapeutic" blood level. Once achieved, this blood level must be sustained by regular dosing in order for the drug to remain effective in preventing anxiety attacks.

Some psychiatrists don't suggest medications because they fear their patient may become dependent on the drug. In my experience, simply making a medication available to the anxiety sufferer leads to a much more benign and helpful sequence, as described by one of my own patients.

"When I discovered that the drug could manage my anxiety, I carried it with me everywhere. Whenever I felt anxious, I took the pill. After a while, I didn't actually have to take the pill, but could feel better simply by reminding myself that I had the pill with me and that I could take it if necessary. As a result of the confidence this thought provided me, I rarely had to pop a pill. Then, one morning, I discovered halfway to work that I had left my pills in my apartment. A few weeks earlier, such an experience in itself would have been enough to set off an anxiety attack. But instead of returning home for the medicine, I just continued on to work. I reasoned that if things got bad I could get through it just knowing that a drug existed that would end the anxiety attack. Besides, if things got really bad, I could call my doctor and ask him to phone in another prescription to the pharmacy down the street from my office."

Instead of creating dependency, the medication enabled my patient to control the most agonizing aspect of anxiety: the perception that he could do nothing about it. As he discovered, it wasn't always necessary to take the medicine. Just knowing that a drug that could end the anxiety attack existed and was available to him proved sufficient.

# 10

# Binge Drinkers and Coffee Nerves
*The Genetic Frontiers of Anxiety*

---

UTURE RESEARCH on anxiety will focus on the genes that provide the chemical blueprints for the neurotransmitters, receptors, and enzymes involved in the action of psychiatric drugs. On the horizon are genetic tests that will guide the diagnosis and treatment of individuals afflicted with anxiety disorders. These tests will also help identify those people most likely to develop an anxiety disorder such as PTSD should they be exposed to traumatic experiences. Such tests will have practical consequences, such as helping in the selection of candidates for career assignments likely to involve emotionally traumatizing situations (Special Forces units come to mind). No predictive test can be 100 percent accurate, of course, since the effects of traumatic stress depend on contributions from both genes and the environment. In addition, mental attitude plays a large role: One person's overwhelming stress is another person's manageable challenge. But the most innovative genetic application will involve the use of tests to help select the best possible drug for each anxious patient. As things now stand, drug selection is largely a hit-and-miss affair. If the initial drug doesn't work, then the doctor chooses another, and perhaps several additional drugs after that, until, eventually, something proves effective.

In addition, drug selection is largely arbitrary and based on the doctor's ability to link the symptoms of his or her current patient with those of earlier patients he or she has encountered with similar symptoms. What worked for the earlier patients should work for the current patient—at least that's the expectation. While such a method involves more than pure guesswork, its effectiveness depends very much on the clinical experience of the treating doctor.

Genetics-driven therapy, in contrast, will be less dependent on the doctor's personal experience. Instead, a given patient's genetic profile will be matched with a drug known to be effective with other patients possessing a similar profile. In a phrase, drug selection based on genetic profiles will replace the approach of the gambler with that of the logician.

"Genetic tests will help guide the treatment for each individual and will, in time, become a foundation for the gradual replacement of our one-size-fits-all medications with new personalized psychiatric drugs," Samuel H. Barondes wrote in his book *Better Than Prozac*.

As a first step in designing personalized antianxiety drugs based on genetic profiles, neuroscientists may have to change their ways of thinking about and classifying anxiety disorders. While the present method based on symptoms can be helpful (e.g., distinguishing PTSD from GAD), it isn't always reliable since, for instance, as mentioned on pages 193–194, anxiety and depression frequently overlap. Neuroscientists are already taking the first steps toward clarification by seeking a genetic explanation for why some people are more anxious than others. Take, for instance, the well-known association between coffee consumption and nervousness, the so-called coffee nerves.

In our coffee-loving society, we collectively consume about 350 million cups of coffee daily. But coffee isn't the drink of choice for everyone. Many people find that coffee exerts an overly stimulating effect on them. After drinking only a cup or two, they feel "wired" and anxious. In other people, the response can be even more extreme. For instance, panic disorder sufferers learn early in their lives that coffee can provoke an attack.

In order to explain this curious association between caffeine, anxiety, and panic disorder, psychopharmacologist Harriet de Wit of the University of Chicago studied ninety-four healthy infrequent coffee drinkers. De Wit and her colleagues collected blood samples from the volunteers and then analyzed the genes that code for two proteins (known as adenosine receptors) that are known to interact with caffeine. One of the adenosine receptors, A2a, is concentrated in the basal ganglia, a cluster of motor regions located deep within the brain. As it turns out, those individuals with two specific variants in the A2a receptor gene experience higher levels of anxiety after drinking coffee compared with people possessing other variations of the gene. This is especially true in people prone to panic attacks.

De Wit's research suggests that since panic disorder sufferers are more likely to have one of the two variants, they should limit their caffeine consumption in order to lessen their chances of inducing a panic attack. Interestingly, it's only their tendency to experience severe anxiety and panic attacks after drinking coffee that distinguishes them. Otherwise, they experience the same basic reactions to caffeine as those without the gene variations (feelings of stimulation, increased heart rate, and faster-than-normal responses on tests of memory).

Of course, it's a long way from finding a gene that helps explain coffee nerves to declaring that individual differences in the experience of everyday anxiety can be attributed to genetic factors. Yet there is good reason for making such a conceptual leap.

IN CHAPTER 4, we discussed the anatomy of fear. As you recall, the amygdala is the critical structure responsible for the fear response. But the amygdala carries out many other duties and therefore shouldn't be thought of as a fear *center*, but rather as one of several components of a fear *circuit*, which involves other parts of the limbic system. In addition, not everyone responds with fear to the same situations; indeed, as pointed out in chapter 8, some people experience little fear or anxiety under conditions that would drive others into fantods. While neuroscientists have long suspected a genetic basis for these differences, they hadn't a clue about the identity of the gene or genes involved.

In December 2002, researchers at the Howard Hughes Medical Institute and Harvard University discovered the gene involved in the synthesis of a protein that acts by inhibiting the fear-learning circuit. The path leading up to this discovery provides an elegant illustration of how researchers achieve new insights by building on earlier discoveries.

First, the researchers set out to discover a gene important in the fear response. In order to increase their odds of finding such a gene, they applied the famous dictum of bank robber Willie Sutton: If it's money you're after, then go where the money is. In this analogy, money in the bank corresponds to the fear response in the amygdala. Thus, if you're looking for a fear gene, narrow your search to those genes most highly expressed in the amygdala.

One candidate gene known as Grp encodes a short protein with a quirky name: gastrin-releasing peptide (GRP). GRP activates receptors on a population of neurons called interneurons located in the amygdala. These interneurons produce the inhibitory neurotransmitter GABA, which decreases the firing of neighboring neurons. When GRP binds to these inhibitory interneurons, it activates them, leading to the production of more GABA and an enhancement of the inhibitory effect on neurons in neighboring circuits. But what if, neuroscientists wondered, the influence of the Grp gene was muted or done away with altogether? What would happen then?

To find out, neuroscientist Eric Kandel and his team at Columbia University bred a colony of mice that lacked the receptor for Grp. In these mice, the interneurons produced less GABA, and as a result the neighboring neurons within the amygdala fired more frequently. As a consequence of this increased neuronal firing, synaptic connections within the amygdala increased in strength. At this point, Kandel conditioned the mice to associate the sound of a neutral tone with an electric shock.

After conditioning, the mice lacking the Grp gene reacted more fearfully than mice possessing the gene: In response to the tone, the genetically modified animals "froze" in place and held that posture longer than the normal mice. What's more, the response was specific to the conditioned (learned) fear. In other potentially fearful situations, such as confronting an intruder, the Grp-deficient mice were no more fearful than normal mice. Only learned fear was enhanced. And remember, all of this resulted from influencing the expression of a single gene!

So far, neuroscientists haven't come up with an "anxiety gene" responsible for all types of anxiety. Probably the closest thing to such a finding is the gene referred to as SLC6A4. It plays a central role in the transmission of the neurotransmitter serotonin.

Each of us inherits either a long or a short version of the SLC6A4 gene. After mixing and matching genes from both parents, an individual may inherit two copies of the long version, two of the short, or one of each. Practical considerations flow from the particular versions of the SLC6A4 gene that we inherit. The amount of the neurotransmitter serotonin ferried across the synapse from one nerve cell to another varies as a consequence of whether we inherit either the short or the long version of the gene. The short version transports serotonin less efficiently. As a consequence of less-efficient serotonin transport, individuals with one or two of the short versions of the gene show abnormal levels of anxiety.

In a test of the effects of the SLC6A4 gene on anxiety, volunteers looked at a series of pictures depicting faces with angry or frightened expressions. Individuals with the short form of the gene showed greater activation of the amygdala on the right side of their brain, which is specialized for processing faces. This could explain why individuals with this gene develop anxiety when encountering faces that don't arouse an emotional response in individuals lacking the gene.

The SLC6A4 gene is also turning out to be an important determiner of a person's tendency to abuse alcohol. In a survey of college drinking patterns, those students with two short forms of the gene (two s's) were more likely to drink excessively and twice as likely as others to be in the top ranks of binge drinkers—defined as engaging in ten or more heavy drinking sprees in the previous two weeks. Nor should this correlation come as a surprise. Alcohol abuse frequently coexists with anxiety. Initially, the alcohol in small amounts serves as a tranquilizer; subsequent heavy use leads to the development of even greater levels of anxiety and, eventually, alcoholism.

But one must be cautious in interpreting this early gene research on anxiety. Since only a small number of people were included in the studies, one isn't justified in suggesting that neuroscientists have discovered an anxiety gene. Instead, it's safer to consider genes as only one factor among others leading to an overly anxious response. Moreover, the consequences of an overly anxious disposition may vary depending on the circumstances.

"This inherited reaction to potential danger may be a boon or a bane," according to Daniel Weinberger, the lead researcher on the study. "It can place us at risk for an anxiety disorder, or in another situation it may provide an adaptive, positive attribute, such as increased vigilance. We have to remember that anxiety is a complicated multidimensional characteristic of human experience and cannot be predicted by any form of a single gene."

Nor can anxiety be pinned on genes alone. Neurotransmitters are also important. Of particular interest is the mood and anxiety regulator serotonin. Researchers at Columbia University have created a genetically modified colony of mice in which they can switch on and off the production of one type of serotonin receptor, the 1A receptor, by giving the mice an antibiotic. In adult mice, turning off the receptors doesn't change the animal's anxiety level. But among mice given the antibiotic while still in the womb, the shutdown results in anxious mice that remain anxious thereafter, even if the researchers switch the receptors back on when the mice reach maturity. Such findings suggest that anxiety circuits are especially vulnerable during the early stages of development.

Anxiety's multidimensional character is especially obvious in regard to anxiety that arises from specific experiences. Consider this rather disturbing fact: Under certain circumstances, any of us can develop a conditioned fear response. If you're involved in an auto accident, or encounter a sufficient number of close calls while driving, you may start to experience a conditioned fear response whenever you hear the screech of brakes. If this experience occurs often enough, the conditioned fear may become so overpowering that you develop a phobia about getting into a car. But with treatment, or even just the passage of time, your anxiety would most likely eventually abate.

But if you inherited the short form of the SLC6A4 gene, your response may be even more dramatic: You could experience an overpowering panic response after only a minor fender bender or a single near accident. Further, your anxiety may spread to situations only marginally related to actual car travel. You may start monitoring newspaper and television sources in search of reports of auto accidents; or you may begin talking about accidents with anyone willing to engage with you in such discussions. If matters progress even further, auto accidents may eventually become an obsession that occupies your mind during every waking moment. Of course, this more severe degree of anxiety would also likely respond to medications and in time remit.

As you have no doubt observed from my descriptions of the research described above, serotonin plays a major role in the production of anxiety. People with low levels of serotonin are prone both to excessive anxiety and exaggerated, often inappropriate displays of aggression. And while animals also become more aggressive in response to lowered brain serotonin activity, do they necessarily become more anxious?

In order to find out, researchers at Case Western Reserve University developed a mouse missing a gene, Pet-1, which is active only in serotonin brain cells. The so-called knockout mouse (a reference to the fact that a gene has literally been knocked out) is normal in every respect except for the absence of the Pet-1 gene.

Deprived of the Pet-1 gene, the mouse's brain fails to develop normal levels of fetal serotonin neurons; moreover, those mice that nonetheless manage somehow to develop near-normal serotonin levels still behave abnormally. Such a mouse will promptly attack another mouse placed in its cage—behavior that is strikingly different from that of a normal mouse. Typically, mice are curious about intruders and attack only after

some preliminary attempts to get to know the unexpected visitor. The Pet-1-deprived mouse, in sharp contrast, doesn't leave
any room for doubt about its aggressive intentions.

So much for aggression. Determining whether the knockout
mouse is also more anxious requires a somewhat subtler
approach. A team of researchers, led by Evan Deneris, placed
a genetically modified mouse on a T-shaped platform located
several feet from the ground. One arm of the T resembled an
enclosed tunnel, while the other arm was an exposed plank. A
normal mouse typically explores both arms of the platform,
moving without signs of distress from the protected to the
exposed arm. A Pet-1-deprived mouse, in contrast, spends all
of its time within the tunnel of the protected arm—the correlate, according to Deneris, of "anxiety-like behavior."

Keep in mind that the human and mouse serotonin systems
share many anatomical and functional features, as well as the
same Pet-1 gene. And although no one is proposing the creation of a Pet-1-deprived human, the knockout research in
mice is the first animal model that may provide a greater
understanding of the causes of abnormal anxiety, along with
new drugs to treat it.

As noted in chapter 9, antidepressant drugs like Prozac and
Paxil work by increasing serotonin activity in the brain and
thereby inducing calmness by modulating anxiety and aggression. The hope is that the knockout mice may prove helpful in
identifying people prone to excessive anxiety.

"If genetic variants of Pet-1 are associated with excessive
anxiety in humans, then tests to detect these variants might be
useful for early diagnosis of people who may be at risk for
developing anxiety," said Deneris. It should also prove helpful

in screening new and even more specific and effective anti-anxiety drugs.

Perhaps about now you're wondering why I'm subjecting you to this degree of detail about the outer frontiers of genetic research in anxiety. My point here is that people differ in their susceptibility to developing anxiety not because of "neurosis" or a "weak will" but, we're learning, because of genetic susceptibility: Nature dealt them a "bad hand."

Under conditions that wouldn't make most people all *that* anxious, the genetically susceptible person may experience heightened anxiety and—if the anxiety becomes sufficiently severe—the development of one of the anxiety disorders we discussed in chapter 7. I think it's helpful to ponder that fact if you have to deal on a regular basis with an anxious person. It's even more important to ponder it if you happen to be an anxious person yourself.

Based on recent research findings, we can now begin thinking of anxiety the way we think of weight, height, and temperament: Genetics plays an important, though not exclusive, role in the final determination. All of which leads to a thorny question: If a person's tendency to develop anxiety is at least partly genetic, should those people with an increased risk be given medicine early in life, prior to the onset of anxiety experiences?

Imagine the following situation. At birth, an infant is found to carry an abnormal variant of the Pet-1 gene. Over the next year or two of life, he or she behaves in conformity with Jerry Kagan's "inhibited" child as described in chapter 3. This combination of abnormal genotype and inhibited behavior makes it likely, although no one can determine *how* likely, that the child will grow up to be an anxious adolescent and adult. Now, if you were the parent of such a child, would you want him or her placed on an antidepressant drug aimed at correcting serotonin levels?

Before deciding how you'd respond to that question, you would probably want answers to several other questions. Would the serotonin drug work as intended and decrease the anxiety, or would such a drug eliminate anxiety altogether? And if it did that, what would be the long-term consequences of remaining on such a drug over an entire lifetime? I expect by now you can guess the position I would take. As I have argued throughout this book, anxiety is a normal response that we subdue at our peril. Without anxiety, we wouldn't be fully human. We want, therefore, to retain our capacity for experiencing anxiety whenever we're at risk, either as a consequence of our own behavior or secondary to circumstances over which we can exert some measure of control.

Anxiety isn't something to be eliminated; rather, it should merely be controlled, lest it become the dominant influence on our behavior. So since we can't eliminate anxiety, we should look for ways of dealing with it. Therefore, as a conclusion to our exploration of anxiety, let's get practical. What can you do to manage your own anxiety?

# Epilogue

## A BLUEPRINT FOR TRANSFORMING YOUR ANXIETY

*Try to Think Critically About the Things That Are Making You Anxious*

Your goal should be to learn as much as possible about your own personal anxiety "themes." Since these themes vary from one person to another, each of us becomes anxious about different things. In general, our anxiety and our sense of competence are inversely related: We become anxious in response to our self-assessment that we lack competence in some area and, as a result, may fail in some conspicuous way and lose face in front of others. Try to identify the common anxiety themes within your own life by thinking back to times when you've felt the most anxious. Anxiety can thus serve as an indicator of those aspects of your life you should try to improve upon. Take specific steps to bolster your confidence in those areas you are anxious about. If you're constantly worried about finances, consider your anxiety as a challenge to learn more about budgeting, look for a better paying job, or perhaps consult a financial consultant. Seek approaches that will help resolve some of the uncertainty that may be fueling your anxiety.

*As an Aid in Carrying Out the Previous Suggestion,*
*Start a Journal*

Write down the specific thoughts that assail you during those moments when you're anxious. Prior to managing your anxiety, you must first recognize it when it arises and describe it. A journal can help with both goals. When you're anxious, turn to your journal and describe the experience. While such an exercise might initially seem likely to *increase* your anxiety, just the opposite usually occurs. Think of anxiety as primarily a background event, like the "scary" music in a horror movie. The music contributes greatly to making the scenes in the movie frightening. If you want to test this for yourself, turn off the sound while watching a video of a horror movie. You'll notice a decrease in tension—the scenes don't arouse the same suspense. Now turn the sound back on, and this time pay minimal attention to the scenes on the screen and concentrate on the music: Shift it from the background to the foreground of your consciousness so that you're now primarily listening to the music and only responding to the story as background. You'll notice that this background-foreground shift also decreases the suspense. A similar process occurs when you move anxiety to the center of your consciousness and describe it in your journal. It thus becomes objectified and available for examination, instead of being just a nameless feeling.

Working on a *written* description is very important, incidentally. The process doesn't work if you merely focus on your anxious experience but fail to describe it in writing. Writing down your experience transforms a feeling (an emotion) into words and thereby deprives that feeling of its privileged access and control. The simple act of writing a description of anxiety—both the content (what comes to mind as a possible cause for your anxiety) and how anxiety actually *feels*—decreases its intensity. Describing your anxiety in writing also works better than speaking about it, such as by dictating your thoughts into a recorder and then listening to your narrative description. That's because listening to your description of the anxiety episode, as opposed to writing and later reading about it, exposes you to many of the autonomic accompaniments of anxiety: the tension in your voice, your slightly increased breathing rate, and so on. In order for this to be helpful, you would have to consciously inhibit the spoken expression of your feelings. Airline cabin attendants are well aware of this. That's why onboard announcements are always given in a calm, composed tone of voice even under emergency conditions. But while such an approach may be desirable to the cabin attendants, it isn't likely to help you manage your anxiety. Therefore, rather than dictating a narrative, describe your anxiety in writing. That way you are better able to use the logical verbal processing of the prefrontal lobes to override the emotional influence of the preverbal amygdala.

*Between Anxiety Episodes, Stop Concentrating on*
*How You're Feeling—Thereby Endowing Your Feelings*
*with the Power to Determine Your Behavior*
Think back on your own life experience. What has made you feel good about yourself and your life? Those occasions when

you sat around anxiously deciding what you *felt* like doing? Or the times when you immersed yourself in a project or accomplished a goal? Most likely, your feelings of satisfaction, joy, and confidence resulted from something you *did*. That's because what you did was under your control, and therefore you could justifiably take credit for it. How you felt, on the other hand—your moods, impulses, and inclinations—weren't controllable. Rather, they were part of a general emotional background in which you could fruitlessly immerse yourself ("How do I feel about this assignment?"), or which you could ignore and simply get on with your project. Take the same attitude toward anxiety: Focus on what you're doing at the moment no matter how inconsequential it may seem and let your feelings sink into the background. As help in doing this, here's a way of using your journal to compare and contrast your activities and your feelings at various times during the day.

Write down on the left-hand side of the page your mood, thoughts, and inclinations: what you would like to be doing if you could do as you wished. On the opposite side, write down where you are and what you were actually doing immediately prior to making the entry in your journal. After only a few entries, you'll notice that if you're like most people, your feelings and activities are frequently out of sync: You do things because they have to be done, not because you feel like doing them. Even during our free time, most of us are often doing one thing when we'd like to be doing something else. Most successful people accept and learn to work with this disconnect between activities and feelings over their lifetimes. They prioritize in such situations, assigning a lower priority to their feelings in the interests of getting on with the task at hand.

Certainly, a person who goes to work only when he or she "feels like it" isn't likely to hold a job for very long. In order to be successful and effective, moods and inclinations can only be permitted to play minor roles in determining workday and leisure activities. One of the ways of bringing about this inner adjustment is by practicing this journal exercise for several weeks. By doing so, you'll lessen the influence of anxiety by shifting your focus toward what you're doing, rather than how you're feeling.

Incidentally, this suggestion isn't in conflict with the previous one that you describe your anxiety episodes in your journal. Journal writing provides a means of preventing your feelings from dictating your life. Once you establish the journal habit, you're less likely to automatically allow anxiety to take control. While this habit won't eliminate anxiety from your life, it will help you to manage it. So when you're feeling sufficiently anxious that you're distracted from whatever you should be doing, write down how you're feeling. That act alone will redirect you toward getting on with desired or necessary activities.

*Learn to Repress Your Thoughts Whenever You Find Yourself*
*Ruminating or Constructing Disturbing Internal Scenarios*
Anxiety feeds on our internal fantasies in which we rehearse in vivid detail all of the things than can possibly go wrong. When you find yourself lost in such fantasies, override them. Try to avoid the tendency to "catastrophize" everything. Put the subjects you're anxious about into perspective. The easiest way of doing this is by simply putting them out of your mind. Say to yourself, "I'll think about it tomorrow," as Scarlett

O'Hara put it in *Gone with the Wind*. Such attempts at actively eliminating from your mind anxious thoughts and fantasies isn't harmful, incidentally, as indicated by some recent research by Karni Ginzburg, a faculty member of the Bob Shapell School of Social Work at Tel Aviv University in Israel.

In 2002, Ginzburg studied more than a hundred patients hospitalized for a heart attack. Not surprisingly, most of them were experiencing anxiety. After their heart attacks, Ginzburg questioned the patients about symptoms of acute stress: trauma flashbacks, difficulty concentrating, insomnia, and irritability. Seven months later he reevaluated the patients during follow-up visits to their homes. His goal was to determine the health of those patients with repressive coping styles marked by attempts to banish from their consciousness all anxious thoughts about their trauma, even in the face of increased heart rate, elevated blood pressure, or other expressions of physiological anxiety they might be experiencing.

Patients who used repression (essentially denial) as a coping mechanism had lower rates of acute stress disorders than highly anxious patients, but had higher rates than patients rated low in overall anxiety.

"Repressors tend to perceive themselves as competent, self-controlled, and having adequate coping skills," said Ginzburg. "The repressor is able to approach trauma-induced emotions and thoughts gradually, in small doses, and without being overwhelmed by them, and also to maintain his or her hope and courage."

If Ginzburg is correct, a repressive coping style is not an inauthentic way of shielding oneself from the totality of an anxiety-inducing experience, but an effective coping style under circumstances that threaten to overwhelm you when you're anxious.

*Adopt Regular Routines and Try Not to Vary from Them,*
*Especially When You're Feeling Anxious*
At the very least, schedule regular times for meals and sleep. But remember: Regular routines aren't the same as rituals. Your goal is to take control of your anxiety, rather than allowing it to control you. How do you distinguish a healthy routine from a ritual? Easy. A routine allows for exceptions, whereas any deviation from a ritual results in increased anxiety. In contrast to a ritual, a healthy routine doesn't force you to conform lest you become anxious; you may even feel less anxious when breaking your routine (occasionally walking or bicycling to work rather than driving or taking public transportation).

*Stick to Your Priorities During Times of Heightened Anxiety*
Determine each morning what you want to accomplish that day and stick to your goals no matter how anxious you feel. If you do this, your brain will soon learn to operate at near-optimal levels under conditions of anxiety. Remember: The brain is constantly renewing itself and changing its "programming" based on your thoughts, attitudes, and actions. So if you want to reduce the level of anxiety that you're feeling, act as if you've already accomplished your goal.

*Avoid too Much Free Time*
Free time is the breeding ground for anxiety. The more time you have on your hands, the more likely you are to start ruminating about something. Some people address this challenge

by becoming workaholics. But, paradoxically, overwork often increases rather than decreases anxiety because of the resulting fatigue, sleeplessness, inability to relax, and self-imposed isolation. "Compulsive work is perhaps the most common way in America of allaying anxiety," wrote Rollo May in *The Meaning of Anxiety*. "Although on occasion this may be a sound reaction to anxiety—work *is* one of the handiest ways of relieving anxiety—it can easily become compulsive." May suggested, "We have to keep in mind the crucial difference between constructive and destructive methods of allaying anxiety." So instead of filling your empty time with an increased workload, take up a hobby or a sport, especially one that brings you into contact with other people.

When you're feeling anxious, remind yourself that physical and mental inactivity only makes anxiety worse. Keeping one's mind occupied, in contrast, immunizes it against the tendency to worry. And worry is not only the most prominent symptom of excessive anxiety; it also leads to rumination. This tendency is at its worst under conditions of physical and mental inactivity. When you're caught up in a cycle of inactivity and anxious worry, ask yourself this simple question: "Can I really do anything to change what I'm worrying about?" If the answer is no, then mentally move on. If your answer is "Yes, I can do something to influence that," then follow up with these two important questions: "*Should* I do anything to bring about a change?" and "Is bringing about that change the best use right now of my time and resources?" If the answer to either of those two questions is "No" or "I'm not sure," direct your attention away from your anxious worrying and turn to something else.

*Try to Maintain Perspective*

In study after study, people with religious beliefs and affiliations reported less anxiety than their irreligious counterparts. If you were raised in a religious tradition, gain perspective by renewing or intensifying your involvement with the relevant practices. If you're not religious, then turn to other means of getting your attention off yourself and combating narcissistic concerns. Reading books on astronomy, archaeology, geology, or similar subjects dealing with the Big Questions will provide perspective and help keep you grounded and less preoccupied with temporary, purely personal pursuits and interests. When you're anxious, you're focused on internal events, on the narcissistic world inside your head.

*Turn to Family, Friends, or Colleagues When You're Feeling Anxious*

Express your concerns and give others a chance to respond. But remind yourself that you're not engaging in a therapy session but a conversation. Sometimes you'll find that the other person is anxious about the same things. Sometimes it isn't even necessary to talk: Simply being with someone can prove effective in reducing your anxiety. When I'm feeling anxious, I often call up a friend and invite him or her to browse with me in a bookstore, followed by lunch or afternoon tea. Isolation increases our anxiety, because when we're alone our fantasies escape the correction provided by another person's perspective.

*Make Some Form of Exercise a Regular Part of Your Routine*

Anxiety includes a large motor component: When we're anxious, we have a hard time sitting still, feel confined, and engage in various "nervous" gestures and movements. Exercise provides an outlet for that inner sense of restlessness. The best exercises are those that bring you together with other people, like golf or tennis. But when you have to go it alone, walking or swimming relieves tension, provides a break from anxious fretting, and improves your physical and—according to current research—your mental health as well.

## SOME FINAL THOUGHTS

Throughout this book, we have examined anxiety from many different perspectives, ranging from molecules to societies. As a result of our inquiry, certain conclusions about anxiety seem reasonable.

First, anxiety is part of our genetic makeup. We wouldn't be alive today if our ancestors lacked the ability to imaginatively anticipate dangers and threats. So accept the fact that since anxiety is hardwired into your brain circuitry, you cannot escape it or deny it. Anxiety is as natural a part of our being as breathing, eating, or sleeping.

Second, anxiety is tightly interlinked with our enhanced powers of thinking, envisioning, and feeling. "The more original a human being is, the deeper is his anxiety," wrote Søren Kierkegaard. With the development of the frontal lobes, our species achieved a quantum leap when it came to enhancing our originality and envisioning future possibilities. As a result, we aren't tethered to the here and now, but can imaginatively anticipate the good things that might happen to us. But we can also envision the bad things—and, as a result, experience anxiety. We can't have one without the other. Anxiety therefore isn't something to be eliminated, but rather to be controlled. It isn't undesirable unless it becomes extreme.

At optimal levels, anxiety functions as an early warning sign of impending danger. And by heeding rather than ignoring our anxiety, we can save ourselves a good deal of trouble in life; dampening or eliminating normal anxiety, in contrast, decreases our chances of coping with potentially threatening situations. So when you or someone close to you experiences anxiety, remember that anxiety provides the stimulus for responding to survival threats. Indeed, the absence of anxiety can prove a costly liability.

At the same time, too much anxiety, or anxiety that is out of proportion to the potential threat, can exact a heavy toll. Most of us do our best when we're at the midway point along the continuum extending from excessive anxiety at one end to no anxiety at all on the other. (Extreme athletes and other inveterate high-risk takers are the exception, as we learned in chapter 8.) In order to perform at our best, we should aim at remaining within the "zone" of optimal alertness and optimum anxiety: not too tired, not too wired.

Finally, anxiety is not only a necessary component to normal psychological functioning, but it also provides a rough-and-ready measure of our capacity for feeling and expressing all the other emotions. "Our capacity for dealing with symbols and meanings, and for changing behavior on the basis of these processes—all are processes which are intertwined with our capacity to experience anxiety," wrote Rollo May.

Contemporary poet C. K. Williams captures the danger of the anxiety-free life in his poem "Chaos":

*. . . the prospect of living without any anxiety renders us*
*even more anxious, ever more ready to accede*
*to beliefs and interests contradicting our own.*

Think of anxiety, therefore, not as a burden to be eliminated
at all costs but as a stimulus for greater accomplishment and
enhanced self-knowledge. You will function at your best when
you accept and attempt to transform your anxiety rather than
deny it or try to escape from it. Indeed, to deny or flee from
anxiety is to renounce your very identity and being. In the
words of philosopher-psychiatrist Karl Jaspers: "Large numbers
of modern people seem to live fearlessly because they lack
imagination. They suffer from an impoverishment of the heart.
Total freedom from anxiety is the inner expression of a profound loss of personal freedom."

# Bibliography

**BOOKS**

Glassner, Barry. *The Culture of Fear*. New York: Basic Books, 1999.

Kierkegaard, Søren. *The Concept of Anxiety*. Ed. and tran. Reidar Thomte. Princeton, N.J.: Princeton University Press, 1981.

May, Rollo. *The Meaning of Anxiety*. New York: W. W. Norton, 1977.

Richard J. McNally. *Remembering Trauma*. Cambridge, Mass.: Belknap Press of Harvard University Press, 2003.

Schneier, Bruce. *Beyond Fear: Thinking Sensibly About Security in an Uncertain World*. New York: Copernicus Books, 2003.

Stein, Dan, J., and Eric Hollander, eds. *The American Psychiatric Publishing Textbook of Anxiety Disorders*. Arlington, Va.: American Psychiatric Publishing, 2002.

**JOURNALS, MAGAZINES, AND NEWSPAPERS**

Davis, Michael. "Are Different Parts of the Extended Amygdala Involved in Fear Versus Anxiety?" *Biological Psychiatry* 44 (1998).

den Dekker, Guido, et al. "From Human Insecurity to International Armed Conflict." Paper presented at the Fiftieth Pugwash Conference on Science and World Affairs, "Eliminating the Causes of War," Cambridge, Eng., August 2000.

Dionne, E. J., Jr. "Inevitably the Politics of Terror: Fear Has Become Part of Washington's Power Struggle." *Washington Post,* May 25, 2003.

Dolan, R. J. "Emotion, Cognition, and Behavior." *Science* 298 (November 8, 2002).

Eisenberger, Naomi I., et al. "Does Rejection Hurt? An fMRI Study of Social Exclusion." *Science* 302 (October 10, 2003).

Helmuth, Laura. "Fear and Trembling in the Amygdala." *Science* 300 (April 25, 2003).

"Learning to Feel Safe: Extinction of Conditioned Fear." Symposium at the annual meeting of the Society for Neuroscience, New Orleans, La., November 2003.

"The Logic of Irrational Fear." *Economist* (October 19, 2002).

Merikangas, Kathleen Ries, et al. "Vulnerability Factors Among Children at Risk for Anxiety Disorders." *Biological Psychiatry* 46 (1999).

Noldus, Lucas P. J. J., et al. "EthoVision: A Versatile Video Tracking System for Automation of Behavioral Experiments." *Behavior Research Methods, Instruments, and Computers* 33 (2001).

Panksepp, Jaak. "Can Anthropomorphic Analyses of Separation Cries in Other Animals Inform Us About the Emotional Nature of Social Loss in Humans?" *Psychological Review* 110 (2003).

———. "The Emerging Neuroscience of Fear and Anxiety." In press.

———. "Feeling the Pain of Social Loss." *Science* 302 (October 10, 2003).

Perrin, Wendy. "Calculating the Odds." *Condé Nast Traveler* (February 2003).

Schwartz, Carl E., et al. "Inhibited and Uninhibited Infants 'Grown Up': Adult Amygdalar Response to Novelty." *Science* 300 (June 20, 2003).

Spencer, Jane, and Cynthia Crossen. "Why Do Americans Feel That Danger Lurks Everywhere?" *Wall Street Journal*, April 24, 2003.

Spielberger, C. D., and R. L. Rickman. "Assessment of State and Trait Anxiety." In *Anxiety: Psychobiological and Clinical Perspectives*, ed. Norman Sartorius et al. London: Taylor & Francis, 1991.

Taylor, Mark C. "Awe and Anxiety." *Los Angeles Times*, September 28, 2001.

Thornton, Jim. "The Joy of Fear." *National Geographic Adventure* (June–July 2002).

# Index

# About the Author

Richard Restak, M.D., a neurologist and neuropsychiatrist, is the author of fourteen previous books on the human brain, including the bestsellers *The Brain, Mozart's Brain and the Fighter Pilot,* and, most recently, *The New Brain.* A Clinical Professor of Neurology at George Washington University School of Medicine and Health Sciences, he will serve in 2005–2006 as president of the American Neuropsychiatric Association.